THE NATURAL HISTORY OF HIDDEN ANIMALS

Other works by Bernard Heuvelmans

On the Track of Unknown Animals
The Kraken and the Colossal Octopus

both published by Kegan Paul

A modified version of the monster of Hans Egede, after the Naturalist's Library of 1839.
Previous and facing page: Views of the mola-mola according to Miss S. Lowell, 1890.

Bernard Heuvelmans'

THE NATURAL HISTORY OF HIDDEN ANIMALS

Edited with an Introduction by
Peter Gwynvay Hopkins

Routledge
Taylor & Francis Group
LONDON AND NEW YORK

First published 2007 by
Kegan Paul International

2 Park Square, Milton Park, Abingdon, Oxon OX14 4RN
711Third Avenue, New York, NY 10017, USA

Routledge is an imprint of the Taylor & Francis Group, an informa business

First issued in paperback 2016

British Library Cataloguing in Publication Data
A catalogue record for this book is available from the British Library

ISBN 978-1-138-97678-8 (pbk)
ISBN 978-0-7103-1333-1 (hbk)

Publisher's Note
The publisher has gone to great lengths to ensure the quality of this reprint
but points out that some imperfections in the original copies may be
apparent. The publisher has made every effort to contact original copyright
holders and would welcome correspondence from those they have been
unable to trace.

EDITORIAL NOTE

This work was composed under the direction of the author, Dr Bernard Heuvelmans, President of the International Society of Cryptozoology, before his death in 2001. The contents have been drawn from his various works, including unpublished manuscripts, as well as his scientific articles.

Briefly put, it is based on the enormous body of documentation, for the most part as yet unpublished, which was amassed by him in his Centre for Cryptozoology in the course of nearly a half-century of work in this field.

My involvement with Dr Heuvelmans began in April 1994 when two of his admirers and supporters, senior members of the International Society of Cryptozoology, arrived at my offices. After a fascinating meeting we arrived at the idea of publishing Dr Heuvelmans' collected works in English, for which a contract was eventually signed by the author. It was an unbelievable task which I hasten to say is far from completed.

We began with the publication of his most important work, a much enlarged edition of *On The Track of Unknown Animals*. A shorter version of this work had been previously published by Rupert Hart Davis, London, in 1958. This was followed in 2003 by *The Kraken and the Colossal Octopus*, an amalgam of two works which had already been published in French.

Both of these titles were a great success, *The Kraken and the Colossal Octopus* coming out just at a time when marine scientists began accepting that there was much more unknown life at the bottom of the oceans than they had first suspected, and they had begun systematically searching and tracking the floors and depths of the sea using the latest scientific instruments.

His supporters all wanted Heuvelmans to sanction the putting together of some of his material, a lot of it unpublished, to create something like 'an introduction to cryptozoology' that was not too long. This is it, as blessed by the master before his death in 2001.

Since that time we have been editing, though not altering, the text and finding appropriate illustrations.

Dr Heuvelmans was quite difficult to communicate with. He lived in a small, somewhat off the track French village. Parcels were sent to the local post office. Occasionally, I could get in touch with him on the telephone and once I had the pleasure of meeting him. I would not use the word 'recluse' but he had to be searched for and, when found, he was delightfully friendly, but not overbearingly so, in the secure knowledge that he was right. Rather similar to his subject matter, really.

The text of this work has been translated by diverse hands over some years. All of them were close to Dr Heuvelmans' thinking and his archives. All translators had English as their mother tongue and the author approved their works before his death.

CONTENTS

Following page: Pontagruel's physter, as drawn by Gustave Doré

EDITOR'S INTRODUCTION

C ryptozoology, to give a precise definition of it, is the scientific study of animal forms, the existence of which is based only on testimonial or circumstantial evidence, or on material proof judged to be insufficient by some.

The word is derived, like so many words in science, from the Greek roots *kryptos*, meaning 'hidden', *zoon*, meaning 'animal' and *logos*, meaning 'discourse': in short, 'the science of hidden animals'. The word appeared in print for the first time in 1959 in a French publication, in English translation entitled *Cynegetic Geography of the World*, by Lucien Blancou, Honorary Inspector of the Game Preserves and Wildlife for the French Government. He dedicated the volume to 'the Master of Cryptozoology'.

The 'Master' was of course Dr Bernard Heuvelmans, the man who invented the subject, although there had been identifiable people before him who hinted at it. As Heuvelmans' research progressed, he found that he began to use the term more and more, particularly after the publication of his seminal work, *On the Track of Unknown Animals*, in French in 1955, in English in 1958, and subsequently in numerous other major languages.

Bernard Heuvelmans, the universally acknowledged 'Father of Cryptozoology', was born on October 10, 1916 in Le Havre in France, of a Dutch mother and a Belgian father. He was raised in Belgium. As a young boy he was very interested in natural history and from the beginning kept all kinds of pets, especially monkeys. He was greatly influenced by the many popular science fiction writers of the time, especially Jules Verne's *Twenty Thousand Leagues Under the Sea* and Sir Arthur Conan Doyle's *The Lost World*.

His higher education was at the Université Libre, Brussels. His doctoral thesis was written on the classification of the hitherto unclassifiable teeth of the aardvark (*Orycteropus afer*), a unique mammal of Africa.

He then spent some years writing about the history of science, publishing numerous scientific articles in the Belgian *Bulletin of the Royal Museum of Natural History*. When the Second World War began he was called up for military service. Eventually captured by the Germans, he escaped four times and subsequently in the chaos after the war made his living as a professional jazz singer, something he had always excelled at, and a popular science writer.

In 1947, he resettled to Le Véginet outside Paris where he made his living as a jazz musician, a comedian, and again as a science writer.

The latent interest in his first love was dramatically sparked into active life again by a sympathetic article in the *Saturday Evening Post* on January 3, 1948, by a famous and well-respected biologist, Ivan T. Sanderson, entitled *There Could be Dinosaurs*.

From 1948 to the end of his life, Heuvelmans relentlessly researched, travelled and collected documentation on 'his hidden world', a collection that grew to a monumental size and which Heuvelmans gave to the Museum of Natural History, Lausanne, Switzerland before his death.

Although Heuvelmans was working on a massive twenty-four volume encyclopaedia of cryptozoology right up until the year before his death, when he became unwell, it was far from finished and it would be a monumental multi-handed task to complete and publish it. Undoubtedly, the original publication in 1955 of *On the Track of Unknown Animals* in French, all 700 pages of it in two volumes initially, and subsequently in many other major languages, particularly English, is the major statement on cryptozoology and the work of its inventor. It is a portmanteau of the field which no one else has even come close to. In this book he made all of his important statements and theories available to everyone.

Ironically, at the time the world became either vastly enthusiastic about the book, its contents and its author or very sceptical, even resorting to ridicule, not that it offended or affected Heuvelmans in the slightest.

However, during his lifetime, particularly in the last ten years of his life, Heuvelmans and we have witnessed a tidal change in the attitude of scientists who believed everything on land and in the sea was already known and classified. The examples of these awkward discoveries are numerous and accelerating on a daily basis.

In 1986, Dr Heuvelmans published a checklist of apparently unknown animals with which cryptozoology should be concerned. He enumerated one hundred and fifty examples, which exceeds considerably the individual cases described in *On the Track of Unknown Animals*. I was very pleased and satisfied to read in the *Quarterly Review of Biology* (Vol. 80, 2005), the bastion of high level science, a favourable review of his book *The Kraken and the Colossal Octopus: In the Wake of Sea-Monsters*. In early years it would never have been allowed into the office of the review editor of that august journal, leave alone get a sympathetic review in its pages.

Reading *On the Track of Unknown Animals* and looking at its body of research one can, to some small extent, appreciate the enthusiasm, audacity, aggressiveness and innovative spirit of the young Heuvelmans. It was a brave stand in the scientific world of the 1950s that dismissed such things as heterodoxy or thinking outside the well-established 'box'.

It took a great many years, but Heuvelmans was accepted in the end and changed scientific research for the better. The work, with age, has changed from new wine to a full-bodied one.

In the last few years of Dr Heuvelmans' life the number of discoveries of 'unknown creatures' became so numerous that only the most 'spectacular' were reported by the world's media. Since his death in 2001, that process has even escalated and will continue to do so. The earth and the sea had not given up all of their 'secrets' by the end of the nineteenth century to be neatly classified. Now everyone realises and accepts this fact.

Sadly, the great man is no more. He died in his bed after being bedridden for a year, alone with his dog at Visnet, which was his Centre for Cryptozoology, on the morning of August 24, 2001 without any suffering, at the age of 84 years.

Heuvelmans had converted to Buddhism in his midlife and was buried in the robes of a Buddhist monk during a private funeral at Le Vesinet on August 27. It was his last wish.

And yet, everywhere across the world, on snowy heights and in torrid jungles, in luxuriant forests and barren deserts, on the sunny seas or in their depths, along tropical rivers and on the shores of secluded lakes or immense swamps, but also in museums, libraries, laboratories and zoos, there is Bernard Heuvelmans or his contemporaries, pursuing for more than forty years his investigations, examining the pieces of evidence placed at his disposal, questioning native eyewitnesses and prestigious scientists, studying the vestiges of the past like the strangest animals of our time, always on the track of unknown animals.

Peter Gwynvay Hopkins
London, March 2007

Postscript

No sooner had I completed writing this editor's introduction on March 15, 2007, than there appeared on the lunchtime television news pictures of a most extraordinary large and stunningly beautiful member of the Pantherinae subfamily, the Bornean Clouded Leopard, which the World Wildlife Fund had just then identified as being a distinct species. I was amazed; Dr Heuvelmans would have been extremely excited and very pleased, but not surprised. P.G.H.

Bornean Clouded Leopard.

WHY CRYPTOZOOLOGY?

'It is said that a foreign scientist is working on a natural history of apocryphal animals. If the title is well justified by the author, he will produce an unusual work indeed.'

(*Revue Britannique*, June 1835)

Throughout the world there has always been interest in mysterious animals - even mysterious men - many of which have become the stuff of legend, and concerning which it is not really known whether they exist or not. This has not changed since ancient times. In those long-ago days, stories were told of dragons, giants and satyrs. Today, we talk about the monster of Loch Ness, the bigfoot of California and the abominable snowman.

In the eyes of the public at large, one thing is sure: it is a fact that the existence of these 'monsters', as they usually are called, is not based on any irrefutable material evidence. Many doubt that such creatures have ever really been seen. In any event, none has ever been captured, for otherwise that would have become known.

So much for what is generally believed and, indeed, endlessly and thoughtlessly repeated. The reality, however, is quite different, and is far more complex and nuanced.

LARGE NEW ANIMALS ARE STILL BEING DISCOVERED

Presently, some 5,000 new animal species are being reported and described each year. And, contrary to what one might think, these are not always minuscule insects or tiny abyssal mollusks.

In 1975, the largest of the peccaries, or wild hogs of the New World, an animal which had been believed extinct, was discovered alive and well in Paraguay. The following year, there was fished up off Hawaii a shark more than four meters long, and belonging to an entirely new genus; it was given the nickname 'megamouth'.

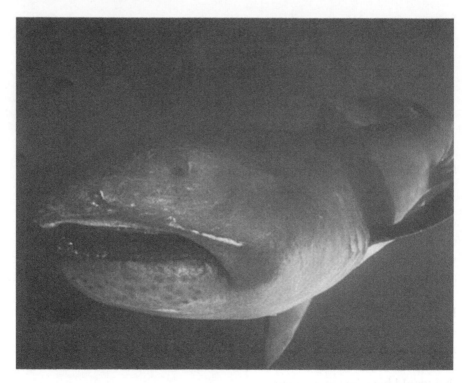

'Megamouth'.

In the course of the past twelve years alone, not only have there been discovered a fistful of birds - all capable of flight, and thus highly visible - but three large reptiles have turned up as well (two giant geckos, one in Iran and the other in New Zealand, and the largest of all the lizards of Arabia, a varan of Yemen). In the realm of mammals *of an appreciable size*, not less than six marsupials were discovered during the same period (an opossum in Colombia, a rock wallaby in Queensland, Australia, a flying opossum, a dasyure and a wallaby in New Guinea, as well as the largest tree-dwelling kangaroo of that vast island), a second warthog of the African deserts, two gazelles, from India and Arabia respectively, a large antelope in Vietnam, resembling an oryx, a muntjac or 'barking deer' of Borneo, and a Malagasy mongoose. Moreover, China alone has given us another muntjac, a pika, i.e., a sort of short-eared rabbit, as well as a civet, to say nothing of a hitherto unknown variety of giant panda. This latter has markings of white and light brown, in contrast to its illustrious brother, which, like the magpie, is plain black and white, and the symbol of fauna threatened with extinction. There has even been discovered a new beaked whale in Peruvian waters, as well as an orca, or killer whale in the Antarctic - altogether, not exactly a skimpy list! But - a fact which has most astonished naturalists - our zoological catalogue has been enriched by not less than eight new species of primates - the order to which man himself belongs: a tarsier, a lemur and a sifaka or propithecus, two marmosets and a lion tamarin, a saimiri or squirrel monkey, and a cercopithecus. Nonetheless, George Gaylord Simpson, the pope of American paleontology, was to declare in 1984, shortly before his death, that it was virtually unthinkable that *even one single* new form of this sort could possibly be discovered.

But the reality is otherwise: let us make the calculation - some forty rather spectacular new species described in twelve years, which comes out to more than three per year! In fact, not a year passes without some surprising discovery being made in the world of rather forthcoming and unwary animals.

CERTAIN MARINE MONSTERS HAVE ALREADY DROPPED THEIR MASKS

From time to time, in the abundant zoological harvest gathered each year, there happens to be unmasked one of these 'monsters' which have figured so largely in story and legend, sometimes since time immemorial.

3

For instance, Virgil (70-19 B.C.), in his *Aeneid*, in mentioning the Mediterranean sea-serpents which came from Tenedos to suffocate Laocoön and his sons, says in his description of them that 'their bloody crests dominated the waves'. Such marine serpents with flaming crests have been described in Northern waters as well. It has been claimed in those regions that such creatures swam ahead of schools of herring and guided them in their migrations, for which reason they were called *Sild Konge* (king of the herring). When a few rare specimens stranded on shore were examined and described scientifically by the Norwegian Peter Ascanius in 1770, and then by his Danish colleague Morten T. Brünnich in 1788, it turned out that they were indeed enormous silvery fish, shaped like ribbons, and whose backbone was decorated, from end to end, with a spiny crest coral-red in colour, and which came to a plume on the head. Capable of exceeding seven meters in length and weighing several hundreds of kilos, these first sea-serpents to be uncovered carry today, in the eyes of science, the name of *Regalecus glesne* or, more commonly, that of 'oarfish'.

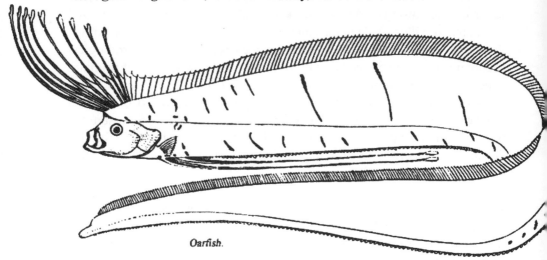

Oarfish.

In 1828, near the Cape of Good Hope, there was harpooned quite by chance a specimen of the largest fish presently known, the whale shark (*Rhineodon typus*), which can reach a length of some twenty meters but which, fortunately for us, feeds exclusively on plankton. Still, this did not prevent one of them from striking terror for several decades in Manila Bay, in the Philippines, where it was described in the press as a voracious sea-serpent and, moreover, spotted like a leopard, clearly another sign of ferocity.

The most gigantic of all rays, the great devilfish, as it is normally called, is in fact another peaceful plankton eater. Ranging up to six meters in size, with a weight of a ton and a half, it was described scientifically in 1829, shortly after the whale shark, under the name of *Manta birostris*. Earlier, along the coasts of Panama and Ecuador, it had been spoken of as a fearful living counterpane, wrapping in its fins any swimmer or small boat within reach, and stifling its victim to death at once. Some forty years after its description, which was in fact quite precise and detailed, a distinguished philologist of South America, Don Enrique Onffroy de Thoron, was to describe this totally inoffensive fish as a monstruous being, because of the two long mobile appendices on either side of its enormous mouth (from which comes the name of *birostris*, i.e., double spur). He took it to be some sort of enormous prehistoric frog, with white arms and classical coaxing siren's calls.

It was only in 1856 that the most unlikely of all marine monsters finally revealed its true nature. This was the kraken, the island-beast of Scandinavian folklore, to which had been ascribed the fearful habit of pulling sailors from their boats by means of numerous slimy tentacles. It proved to be, in fact, a gigantic cephalopod, in other words, an outsize cousin of the octopuses and cuttle-fish, and was given the generic name of *Architeuthis*, meaning 'chief squid'. Certain specimens, which had been found stranded on beaches, exceeded seventeen meters in length and weighed several tons. Nothing astonishing, to be sure, in the fact that the scientific establishment had awaited a veritable epidemic of strandings of this sort in Newfoundland, in the 1870s, before finally recognising the existence of these creatures, after some thirty years of hesitation and dithering!

THE HAIRY COLOSSUS WHICH LIVED UNDER THE EARTH

To be sure, all of the foregoing took place in the boundless depths of the oceans, but similar sensational events also occurred on dry land.

Thousands of years ago, the most ancient Chinese works of natural history recorded - and this observation was still being confirmed by local populations as recently as the last century - that there existed in the far north of Asia, near the ocean, a sort of titanic rodent, as big as a whale or an elephant. It was called sometimes *fen-chu*, or burrowing rat, sometimes

yen-chu, or hidden mouse, and even sometimes *chumou*, or the mother of all mice (the word *chu*, effectively, is used to designate any rodent). However, some people also spoke of it as a species of mole. It was in fact said that this hirsute monster with tiny eyes dug dens under the ice as large as our tunnels, by means of its two huge teeth in the form of pick-axes. It was added that this super-mole with its rodent's incisors died if it happened by accident to reach open air and thus be exposed to daylight. When this happened, the local people collected its enormous teeth, which they then carved into various domestic utensils or decorative objects.

It was not until the end of the eighteenth century that Western science ventured to conclude that this truly incredible beast was none other than the mammoth. Carcasses of this fossil elephant, conserved whole in the frozen peat, sometimes were exposed through erosion and so stood out, in all their furry glory, on the snow-covered tundra. So this is what struck terror into the hearts of roaming trappers, but it never prevented the native peoples from carrying on, since the earliest times, a thriving commerce in the ivory collected from these formidable monsters.

OLD FABLES BECOME REALITY

It was recognised in 1816 that the fabulous *mé* which, according to Chinese and Japanese legends, combined aspects of the bear, the elephant, the rhinoceros, the cow and the tiger, and which was said to feed essentially on iron and copper, was in fact a species of tapir with white hindquarters, living in India. This discovery was all the more striking because this genus of ungulates had been held to be specifically South American. It was, as it should be, entirely herbivorous, but the Oriental fable was nonetheless not entirely without foundation. The Indian tapir is, in fact, black and white, like the giant panda or certain domestic cows. It is equipped with a preliminary shape of elephantine trunk, and it has a profile and eyes which recall those of the Asian rhinoceros. Moreover, it has strong, sharp teeth, which one might well believe capable of tearing metal as well as flesh.

In 1829 the matter surfaced again, but this time on the other side of the world. The pinchaque, a 'hairy phantom' of the folklore of Columbia, said to be as big as a horse, turned out to be a hitherto unknown tapir of the high mountains.

It took Western zoologists ten years to accept, in 1901, the presence in the heart of Africa of a forest zebra, a highly unlikely creature, since these striped horses, with their proverbial speed, are by their very nature swift runners of the savannah. The animal which the pygmies called *atti* or *okapi* turned out on examination to be a giraffe with a shorter neck, and with zebra stripes only on its hindquarters.

And, speaking of pygmies, not just decades but millennia had to pass before their existence was accepted as real. First of all, the question was always raised if those creatures of which the Greek poets spoke - this race of African dwarves, perpetually at war with the cranes - were monkeys or men, or if they were not rather the product of imagination. At the beginning of the last century, the great Cuvier still held them to be a fable, born of the practice with regard to Egyptian monuments of 'representing in the same tableau men of greatly differing sizes: the king or the victor gigantic, and the vanquished or the subjects three or four times smaller'. It was only in 1870 that the Germano-Baltic naturalist Schweinfurth finally had the occasion to see, in flesh and blood, these little reddish and downy men in the region of the Upper Uélé, in the present-day Zaïre. To be quite honest, his discovery was preceded by that made five years earlier, in Gabon, by Paul Du Chaillu, a rather eccentric American of French origin. However, no one wanted to believe him, because he claimed also to have seen there, with his own eyes, gigantic and ferocious apes, similar to men. At the time this was still being passed off as totally improbable, even though the gorilla had been described scientifically in 1852 by Isidore Geoffroy Saint-Hilaire, on the basis of indisputable anatomical elements.

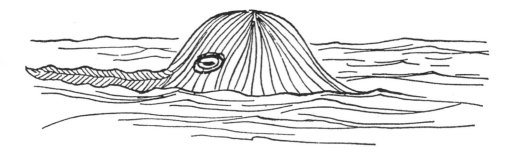

The Robertson sea monster, 1834, after Captain Neill.

7

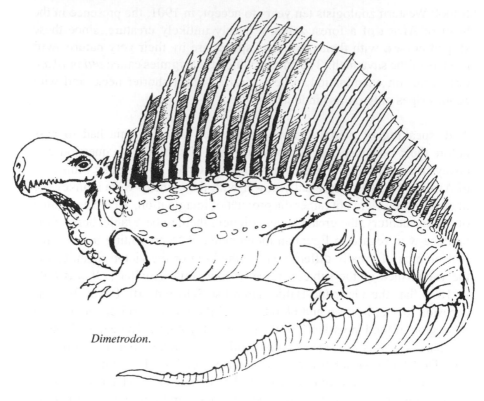

Dimetrodon.

SATYRS AND DRAGONS IN THE 20TH CENTURY

Such examples by no means shook the negative faith of the unbelievers. It was only in 1903 that the truth emerged about the abominable forest-men, hairy giants thirsting for blood and rape, which had been reported for many years in Rwanda. The natives said that they loved women so much that they suffocated them when embracing them. These were the peaceful mountain gorillas, today the stars of an admirable film, *Gorillas in the Mist*, which was inspired by the studious - and by no means improper - life among them of the martyred ethologist Dian Fossey.

In 1912, on the Indonesian island of Komodo, there were at last captured alive dragons more than three meters in length, and weighing between 150 and 200 kilos. More precisely, these were enormous monitor lizards, devourers of macaques, but equally capable of taking on wild boars, deer and buffaloes. The most disquieting rumours had long circulated about these reptiles, which were given the very descriptive name of *Komodo dragons*:

8

they were suspected of anthropophagy, which subsequently proved to be the case. In fact, it appears quite likely that a Swiss baron, Rudolf von Reding, was devoured by them in 1974 (the only thing ever found of him was his camera, doubtless judged too difficult to digest). Further, it is also certain that a young French tourist, despite being an athletic young man, met the same fate in 1977.

It must be emphasised that these stories of innocent satyrs and bloodthirsty dragons are not set in the Olympus of the gods and demigods of ancient Greece, nor do they date in the least from the shadows of the Middle Ages.

THE DESTINY OF FABULOUS BEASTS IS TO BE UNMASKED

Examples of the demythification of legendary animals could be multiplied without end: this is the life-blood of zoology. There is nothing here which stems from the irrational, or from the phantasmagorical or the supernatural, in spite of the sometimes terrifying or burlesque names hung on these alleged 'monsters' in folklore, in superstition or in the most modern media. And all of these phenomena are certainly not to be classified pell-mell in a more than suspect category, just because certain misled souls - hare-brained individuals, practical jokers, hoaxers, hungry for miracles, glory or profit - think that they can establish a correlation between observations, in the USA, of hairy giants with big feet and the passage of UFOs, or between the misdeeds attributed to marine monsters and the fatal trap of the Bermuda Triangle.

Yesterday we saw enter the fold of traditional zoology the sea-serpent with the fiery crest and the one with the leopard spots, the great sea-devil and the kraken, the kidnapper of sailors, the *fen-chu*, the rat of the subterranean ice of Siberia, the metallivorous *mé* and the phantasmagorical pinchaque, the okapi, the impossible zebra of the forest of the Ituri, and the pygmy, which eats its flesh, the salacious forest-man from the shades of the Dark Continent, and the cannibal dragon of the little island of Komodo. Why then would it not be the turn, tomorrow, of the Monster of Loch Ness, the Himalayan snowman and the colossal dragon which haunts the rivers and the marshes of the Congolese basin, and which is suspected to be a surviving dinosaur?

9

THE SCIENCE OF HIDDEN ANIMALS

In the 1950s I alone tried to systematise and accelerate the process of 'discovery', in the proper sense of the term, of such spectacular but elusive animals. After all, this is an operation which is quite normal and even mundane in zoology. This was I, Bernard Heuvelmans, Doctor of Zoological Sciences, and founder of what I was going to call cryptozoology, or 'the science of hidden animals'. These latter are, to a certain extent, still unknown, but they are also ignored, and not without disdain, by conservative circles of science. For this reason they are apocryphal, in other words, they are 'kept secret'. Nevertheless, these are animals like any other, even though their existence is based only on testimonial evidence (a stack of eyewitness reports) or on circumstantial evidence (a range of corroborant indices), or else on material evidence judged insufficient by some.

The essential task of cryptozoology is, first, to establish a physical and behavioural identikit portrait, as precise and detailed as possible, for each apparently unknown animal about which one has significant information and then, if it is truly new, to try to discern its most probable zoological identity. Only then, knowing where, when and how to track it down, can one try with some hope of success to encounter it in nature, in order better to know it and to protect it.

After more than forty years of dogged and uninterrupted research, and after having published on this sort of problem a half-dozen major works, certain of which - translated into some ten languages - have become classics, I finally was able to establish a meaningful balance sheet. There are no fewer than 150 animal forms which are 'hidden', but which nonetheless have been spotted and reported, and which today await being unmasked.

At this hour, when nature is in peril, and when innumerable animal and vegetable species are disappearing, often without our knowledge - more than fifty each day, according to recent estimates - it is well to emphasise by means of striking examples that zoology is not a doctrine frozen in history but, on the contrary, is an adventure of the human spirit, alive and exalting.

2
MONSTERS OF YESTERDAY, TODAY AND TOMORROW

'There is a fictional coloration in everybody's account of an "actual occurrence", and there is at least the lurk somewhere of what is called the "actual" in everybody's yarn.'
(Charles Fort, *Wild Talents*)

T he principal objection generally raised to the real existence of apparently unknown animals is that they have too many fantastic and even supernatural traits to be taken seriously, in other words, to be considered as beings of flesh and blood, and worthy of figuring in zoological catalogues. Scientists with a realistic mentality - and perhaps too much down to earth - and narrow-minded folklorists normally conclude that these manifestly legendary 'monsters' are nothing other than products of the unbridled imagination of mankind.

But, in actual fact, this attitude of rejection, of an *a priori* debunking without any form of investigation, bespeaks not only an ignorance of the history of zoology but also a profound lack of knowledge of the *process of mythification* which affects our entire grasp of the external world and its subsequent reconstitution in our mind.

A STRANGE BIPED SMITTEN WITH INTELLIGENCE

Man is a singular animal. Like all others, he is governed above all by his emotions, but he is beyond doubt the only one which endeavours to pass for an essentially rational being. In this regard, it is significant that the mental images engendered in his senses by the various sensations which he perceives are not in any way to be classified according to categories which conform to a logical system, but rather according to the sentiments or specific emotions which these sensations call forth: pleasure or pain, fear or anger, joy or sadness, heat or cold, attraction or repulsion, appetite or distaste, sexual desire or aversion, admiration or scorn, envy or pity, love or hate, etc. As a result, the various sensory messages - forms and colours, sounds, odours, tastes and tactile impressions - habitually associated with a particular sentiment or emotion become implanted in the same part of the memory and remain henceforth linked there. Thus, to take a number of particularly striking examples, there are associated, on the one hand, the colour red, heat, passion and life, as well as blood, anger, tumult and danger and, on the other hand, the colour green, nature, freshness, calm, but also venom, poison, dissimulation and fear. The relationships sometimes are far from evident. All human beings being constructed of the same identically organised biological units - cells - it is not astonishing that these amalgams, which are fixed and charged with intelligence, prove to be rather similar throughout the entire world. Nevertheless, since environment varies by country, each culture necessarily uses its own imagery, which takes into account its specific differences, which, in any event, are no more than superficial.

IMAGINATION SUFFERS FROM PARANOIA

It is, to be sure, perfectly logical that the very structure of our thinking be tied to the intimate anatomy of our nervous system - which is a sort of miniaturised supercomputer - and to its operation. Ultimately, it is thus that the interplay of constant associations - not in the least fortuitous but nonetheless founded on simple resemblances, analogies and parallelisms - governs our mental processes.

Everything takes place as if each concept had its own family of harmonics. So, this will help us to understand with what facility metaphors and symbols may arise from almost imperceptible shifts, and with what *legerdemain* our daily life can be parodied in a dream, and why those which are called 'primitives' generally bring together all of the objects of the world in a series of classes, each one dominated by a totem, i.e., an emblematic animal, and grouping within itself the most heterogenous things: a division of the tribe, certain stars, animals, plants and stones, a cardinal point, certain materials, an element, a colour, a sound, a particular activity, etc. This also helps to understand how poetic images emerge. Did not Rimbaud, for example, speak of the colour of vowels? And did not Baudelaire say in his sonnet *Correspondances*: '*Les parfums, les couleurs et les sons se répondent*' [perfumes, colours and sounds answer each other]? How judiciously this poem was entitled, for it is a veritable Theory of Correspondences which is to be found at the base of all occultist doctrines, as well as of divinations, or divinatory sciences, and magic, which are the practical applications of it. So, this is how we can explain, in particular, the empirical medicines and pharmacopoeiae linked to the dogma of signatures - of naive analogies - just like the practices of sorcery based on what Frazer, in his *Golden Bough* called 'sympathetic magic'.

This state of affairs also reveals to us by what process of irrational associations a language and, perhaps as a consequence, a writing, develops step by step, and how a simple list of names manages to transmit a message of ethical content. For example, that which at the beginning was supposed to be only a catalogue of all known animals, the *Physiologus*, almost spontaneously became elevated to the moralised *Bestiary* of the Middle Ages. In brief, the constancy of the amalgam in our minds of apparently disparate facts teaches us how, in general, the imaginary operates.

If the imagination merits the sobriquet of 'madwoman of the house', as Malebranche proposed in 1674, its madness then is without any doubt of the paranoid type, for its delirium is rigorously systematised.

THE UNIVERSAL LANGUAGE OF MYTH

A way of thinking in a secretly structured manner thus appears responsible for what one may call 'mythical thought' or, more simply, just 'myth'. The latter, in fact, reflects an archaic, primordial mental faculty, which is nonetheless quite authentic, as it is linked to elementary sentiments and emotions. Ultimately, it can be controlled and disciplined only at a later stage by reason, or 'rational thought', abstracted from reality, which has its base in the cerebral cortex.

In accordance with the process of automatic agglomeration described above, it appears that when information received from the external world traverses the limbic system (that portion of the central nervous system located in the convolution of the hippocampus and sometimes called the 'emotional brain', but in which memorisation also takes place), this information is sorted and stored in the same mental categories. It falls into line with the same stereotypes, to the point of being contaminated and supplemented by them, and, finally, is deformed into the same moulds of the mind. What Carl Gustav Jung subtly termed 'the collective unconscious' is undoubtedly nothing other than an evolutionary adaptation, programmed genetically.

In this case, what are the biological factors which would require the structure of the mind manifesting in the various aspects of mythical thought: the Myth, essentially religious, the Legend, or saga of heroes, and Folklore, in other words, the whole of belief, fables and popular tales? Without doubt it is fear of psychic troubles which can be caused by the vicissitudes of our daily life when they are experienced for the first, and perhaps the only time: the painful experience of birth, the expulsion from the maternal bosom, the icy hostility of the outside world, the momentary deprival of the sensual pleasure linked to the satisfaction of hunger, an all-consuming but sometimes thwarted passion for the mother, the rivalry with the father (this antagonist who hinders suckling around the clock), and then, soon enough, the competition with brothers and sisters, the feeling of abandonment, the first loneliness and, later, unsatisfied sexual desires, combats, defeats, failures and humiliations, deceptions and betrayals, divorces and mourning, and even death itself.

What enables us to view as almost familiar all of these ups and downs, these trials and tribulations, these misadventures - sometimes veritable dramas! - is that they have been inscribed in our genes even before we have been confronted with them. So, this is also why the passage of our existence seems to conform to the same standard scenario, to which the creative mind believes it can add embellishments freely throughout the world, so as to describe the manifestations of forces deified in Nature, in order to recount the magnificent exploits of epic heroes, and indeed fairy tales as well, and even to elaborate on simple funny stories.

In truth, the imagination is strictly limited in its choice of intrigues and motives, and the outlines of its products are pre-planned rigorously.

Olaus Magnus's physter, 1555.

THE UNKNOWN IS HATEFUL

In summary, mythical thought, by its priority and permanence, would serve to protect the representatives of our species against the psychological traumas linked to new experiences, which are all the more distressing because, by definition, they have never before been faced; just as, throughout the ages, our organism gradually developed natural defences to fend off the whole panoply of physical aggressions which are possible.

The science philosopher Léon Brunschvicg said in 1934, 'Primitive people want to explain everything, developed people admit that gaps exist'. This distinction between the thinking of the 'savage' and that of the 'civilised' person appears to us today to be artificial and quite out-of-date. Like it or not, the Unexplained - perhaps the Unexplainable, who knows? - in sum, the Unknown, terrifies everyone. If, to dissipate it, the 'primitive' turns to explanatory myths, the Western savant fills in the gaps in his knowledge by constructing hypotheses, which comes down to the same thing.

So, under these circumstances, what happens in our head when we find ourselves face-to-face with animals which apparently are unknown, or when we have to deal with the problem which they pose for us, and which in fact reflects our ignorance of the inventory of the zoological world? To neutralise the terrifying side of this Unknown, or at least its unpleasant nature, and to reassure ourselves, we are drawn irresistibly to identify the animal in question with one or another of the more familiar forms of our mythical universe, which Jung termed 'archetypes', in choosing, of course, that one among them to which one can most readily identify with.

Fabulous beings are legion; they seem to be almost countless. The incomparable *Dizionario Illustrato dei Mostri* (Illustrated Dictionary of Monsters) of Massimo Izzi, published in 1989, contains several thousands of them. But, if one analyses with care the symbolic significance of the various fantastic creatures so listed and classified, one ends up finding that it is possible to reduce all of them to scarcely more than twenty. This derives from the fact that each one of them appears linked to one or another of our psychological problems, which after all exist only in relatively limited numbers. The 'monsters', as we call them, are in fact the reflection of these conflicts at the surface of our unconscious, and not only of our individual unconscious, but also of our collective unconscious, as revealed by Jung.

Let us review what could be called our 'fundamental monsters', and let us try to discover what each of them, from the outset, can represent for us.

ALL HOPES ARE PERMITTED

The *Chimera*, to start with, this absurd assembly of anatomical fragments borrowed from a diversity of animals, none related in any way to another, is the prototype of all of the apparently impossible beings, whether they be real or fictional. Let us cite at random the *platypus*, a mammal which lays eggs, the horned hare and its various descendants (the numerous *Wolpertinger*, traditionally found in the Germanic countries, and the jackalopes, fabricated in North America), the *chalicotherium*, an ungulate, i.e., one which has hooves, but which is armed with formidable claws, the *basilisk*, a cock which lays eggs and which, moreover, are eggs of crested serpents, the *jumart*, a pretended bastard of horse and cow, and, finally, the endless cohort of all of the paradoxical creatures which flood mythology: white negroes, miniature giants and gigantic dwarfs, leopards without spots and zebras without stripes, flying reptiles and tree-dwelling fish, birds without wings and covered with fur, vegetable-eating carnivores and cannibal herbivores and, let it be said in parentheses (for this does not belong in cryptozoology but in cryptophytology, the science of hidden plants), anthropophagic plants and

Hans Egede's 'most dreadful Monster' after Pastor Bing.

animal-producing trees. All of these contradictory beings deliver a message of hope, since they prove that *everything is truly possible*, including - by implication - our wildest ambitions and our most chimerical dreams.

GOOD, EVIL AND SEX

The *Western Dragon* is the personification of Evil, all that which we have had to combat in order to survive with dignity. The *Eastern Dragon,* on the other hand, is a beneficent creature symbolising authority, strength, experience, wisdom and kindness, all things which we should honour and respect.

The *Unicorn*, phallic symbol *par excellence*, is the image of aggressive virility, of the power of the male, a power which can be subdued by the attributes of a 'weak woman' who appears to be pure and guileless, but is in fact the most perfidious of beings, armed with a deadly efficiency.

17

The *Siren*, on the other hand, is instead the image of the all-enveloping and devouring mother, the Great Lady without mercy, the *femme fatale*, the Vamp, of which the male is eternally the victim. Aquatic and cannibalistic at the same time, she likewise evokes in man the nostalgia for times gone by, when he floated without a care in the world in the amniotic fluid in the folds of the maternal entrails.

The *Amazon* exercises a similar function, but on the social plane. Jealous of the prerogatives of the male, she does not seduce him in order then to devour him, but she ravishes him and then flays him and reduces him to slavery, once his role as breeding stallion has been accomplished.

The *Wild Man* or *Satyr* plays a dual role. On the one hand, he serves as a foil, to make of the man of our times the 'Knight of Old', or to isolate the *chic* and fashionable executive of our times from all which is bestial, uncultivated and repugnant. But, on the other hand, he is the idyllic image of a lost paradise, that of an animality not constrained to work, and not guilt-ridden through the crucifying sentiment of 'sins of the flesh'.

The *Hermaphrodite*, fulfilled to the extreme from the sexual point of view, achieves the gluttonous dream of those who would like to experience the feelings and the sensations of two sexes at the same time.

THE ME AND ALL WHICH MENACES IT

The *Ogre* and the *Ogress*, anthropophagic giants, are a terrifying version of the world of parents and adults, as seen through the eyes of the child. As for the *Little People*, those who live underground, gnomes, goblins and elves, they are only the opposite side of the same myth; a transposition of the universe of children, isolated by their small size, not understood, relegated to an inferior sphere, and obliged to render many services to the 'grown-ups' in order to draw their favour and to be assured of their protection.

The voracious *Bogeyman*, the beast that eats people, is as ambivalent as many other monsters; he incarnates the fear of being devoured, suppressed, and annihilated, but he also represents the desire, and there can be nothing more 'retro', to return to the warm security of the maternal womb.

The *Werewolf* appears to underline the fear and the risk of a brutal regression to the savage animal stage, but its ineluctably tragic destiny incites us not to give way to our 'lower' instincts, which are fortunately usually repressed.

In this regard, the *Lake Monster*, hidden beneath impenetrable troubled waters, evokes all that which is perverse, unavowable and even unspeakable within the depths of our souls. But, as it has the reputation of guarding jealously a treasure, this means that we are profoundly attached to that which we push back into the most secret depths of our hearts.

As for the *Great Sea Serpent*, he is the transparent symbol of the Devil, the Prince of Darkness, in this case of what formerly was called 'the outer darkness', in other words, the immense ocean reaches situated beyond the horizon, and which have always been held as the preferred domain of the powers of 'Evil'.

Just as the *Tentacular Monster*, surging up from the abyss, represents the dangers which threaten us from below, the perpetual conflicts with ourselves, the Kidnapper-Bird, whether it be the Roc bird, the thunderbird, or even the griffon, the latter being more reptilian, is the image of what lies in wait for us on high. The *Demiurge*, the moral authority which judges us and will punish us.

The *Phoenix*, ambassador from a fantastic and distant 'country,' being perpetually reborn from its own ashes, reminds us of the lost Eden, the land where death does not exist. Its symbolic presence reminds us in our imagination of the constant possibility of a return to a past 'Golden Age'.

The makara, sea-monster of Indian legend.

If the *Phoenix* represents the thirst for immortality, the *Vampire* serves above all to warn of the curse which stalks, like a shadow, all those who seek to acquire it.

Finally, there is the long procession of *Different Men*: headless or with two heads, with the face of a dog or a bird, having a single eye or a hundred eyes, deprived of nose or mouth, or again draped in enormous ears, with a single leg or multiple paws, quadrupedal or with feet turned backwards, the lower body that of a he-goat, a donkey or a horse. This interminable list brings to light that from which we have escaped by being what we are, and it thus reassures us in a certain way as to our condition, as miserable as it may appear.

To summarise, there are myths, and mythical creatures, for all ages, all sexes, in fact, for all peculiarities. This is why they attract us, seduce us, trouble us, captivate us, even illuminate us, even though with an obscure clarity, and, finally, reassure us or console us. This is why we are so partial to 'monsters' in the cinema, as well as in painting or in literature.

THE METAMORPHOSIS OF UNKNOWN ANIMALS INTO FABULOUS BEASTS

As paradoxical as it may seem, of all the animals, the fabulous beasts are the closest to us, and are those most closely linked to our intimate life. The dog, the cat, the horse and a few others live at our sides, but as for the monsters, they live in us. They are the most sure guarantors for the peace of our souls.

This is the underlying reason why we are so inclined, and with such impatience, to hang the cast-off clothing of these legendary beasts on animals of flesh and blood, and even sometimes the most prosaic of these. The operation clearly is all the more easy because these latter are poorly known, or only on the threshhold of becoming so.

In this regard, what then are the apparently unknown animals with which cryptozoology is concerned? They are the ones which have been spoken of because they have been named and described by people living in their proximity, or because they have been reported by travellers who claim to have encountered them by chance, or again because persistent rumours circulate about them, or because there exist traditional representations of them. These beings are in fact incompletely known animals.

Still and always this horror of the unknown. Let us seek to fill in the zones of shadow or the clear gaps in knowledge which we have of them by borrowing certain of their missing traits, sometimes, alas, of a fantastic or supernatural nature, from the mythical archetype into which they can be most easily inserted. The less we know of them, either because we have scarcely caught a fugitive glimpse of them, by virtue of the fact that they are aquatic, nocturnal or burrowing, or because they haunt dark equatorial forests, in part flooded, or inhospitable deserts or mountain heights difficult to climb, or again because they fear us, and flee from us, the more they are necessarily called upon to be mythified. It is thus quite natural, and almost ineluctable, to see slightly or poorly known animals soon considered as 'monsters', as creatures of exception, as prodigies, perhaps even as omens. This is irrational. Cicero said '*Monstra appellantur quia monstrant*', implying that monsters show 'the future' or 'that which is hidden'. This was clearly a pun on his part, for this fine linguist knew very well that *monstrum* derives in fact from the verb *moneo*, that is, 'to caution' or 'to warn'. But the lesson remains the same.

MYTHIFICATION RENDERED UNRECOGNISABLE

The veritable metamorphosis which animals yet to be discovered undergo by virtue of mythification can be deep and radical, if the archetypical mould into which one forces them is too cramped and if it resembles rather that most uncomfortable bed of Procrustes. But, however that may be, the treatment in question will henceforth make stick to their bodies, as by decalcomania, the characteristic aspect, sometimes even too extravagant, of the archetype, as well as its often aberrant mores. To rid them of it will always prove to be a long and delicate operation, like the removal of a tattoo.

It may happen that the process of mythification is carried so far that it renders its victim unrecognisable.

Because the manatee has pectoral breasts like its cousin the elephant, and also like a woman, and because its body ends in a fish-tail, it has always been assimilated, on both sides of the Atlantic, as the seductive mermaid with golden hair, and this despite its baldness and its muzzle, hideous to

21

our eyes, and it has at the same time been suspected of the worst and most bloodthirsty instincts. Along the rivers of Central Africa, the manatees are generally described as water vampires, as ghouls capable of draining imprudent bathers of their blood and even their brains, by sucking them out through the nostrils. Now, can one imagine a more inoffensive and more disarming animal than these sorts of aquatic cows, which pass the best part of their time browsing indolently on water hyacinths and other succulent plants?

The original unicorn was - and this is frequently unknown - a man with a single horn in the middle of his forehead. Nevertheless, in the fourth century, an Oriental quadruped similarly armed on its nose was described by a Greek physician by the name of Ctesias. This animal at once came to occupy, in our mythical pantheon, the place of the unicorn man (perhaps the ithyphallic state of the latter was so crudely presented that it could only end up by shocking). Still, the beast in question was none other than the one-horned rhinoceros of India: the text of Ctesias leaves no doubt in this regard. At all events, it found itself mythified to such a point by its illustrators, taking its description to the letter. They made of it a graceful white horse with cloven hooves and with its forehead decorated with a long spear of twisted ivory, imitating the unique tooth of the narwhal. During the Renaissance, the encyclopedists of zoology were not able to identify it and, in their writings, classified the powerfully armoured rhinoceros and the graceful unicorn together, side-by-side.

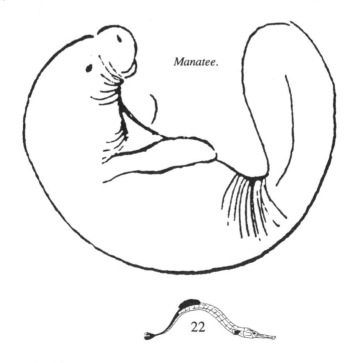

Manatee.

THE ROLE OF THE WILD MAN HAS HAD VARIOUS INTERPRETERS

One given archetypal mould can serve for multiple transformations, either simultanously or in sequence. Many different animals have, at different times, been held to be various types of dragons, devils or sea-serpents, and have been defamed as such in the popular mind in conformity with their traditional reputation and appearance.

The case of the wild man is exemplary in this regard. His image, if one may put it so, has surely been held in the course of prehistory by *Neanderthal man,* who it is now known coexisted, during tens of thousands of years, with *Homo sapiens*. Rougher, hairier, less well armed than the latter, but probably more cunning and more elusive because he had remained closer to his animal roots, the *Neanderthal* doubtless occasionally carried off women of our species, which has contributed to the legend of the satyr.

After he had disappeared little by little from Europe, driven away or perhaps exterminated by his *sapiens* brothers, the first rumours began arriving in the West of the existence in Asia and in Africa of great apes bearing a vague resemblance to man. These proved to be, respectively, the orang-utan and the chimpanzee. It is these last two which, in their turn, were forced into the mould of the wild man, which had become vacant, by calling them straight away forest-men, and suspecting them of refusing to speak in order not to be obliged to work, as well as attributing to them the reputation of being bloodthirsty brutes.

Then, more recently, the lowland gorilla and then the mountain gorilla succeeded them in this role. It was claimed that they stunned elephants with blows of a club - the traditional arm of the wild man in the religious and heraldic art of Europe - and, as usual, they were accused of carrying off black women in order to rape them. We know today that gorillas are especially good-natured and peaceful great apes. They are almost totally vegetarian, and much less obsessed with sex than is Man. Since we have come to know them much better, their bad reputation has been transferred to another anthropoid which is still incognito, and living in the Himalayas. It is this latter which has now been pressed into the mould of the wild man. It is claimed that he knocks out yaks with his bare fists, before gutting them with his teeth and devouring their entrails, and also - need this be said? -

23

tales are told that he attracts women sexually, with a marked preference for little girls. This is the abominable snowman, to call him by his name, which moreover is totally undeserved.

MONSTERS OF THE FUTURE - ALREADY ON STAGE

Let us be reassured, however, that the systematic enterprise of demythification of monsters, which is the task undertaken by cryptozoology, will never lead to the destruction of those beings of which we have such great need. Once the animals which are incarnated in them have rejoined the fold of Science, others will always be found to take over from them and play their role. And if by chance the inventory of the animal world should really approach completion, which the current events of zoology never cease to contradict, we would find elsewhere enough to nourish our elementary myths.

The dark and forbidding monsters which lie in wait for us on all sides have taken on quite another visage. The 'hole in the ozone layer' menaces us from above, the 'pollution by radioactive waste' instills fear from below, like an omen of a new deluge. The devouring crab which is cancer, and an even more treacherous invader, AIDS - these enemies of the interior - have reduced werewolves, vampires and other cannibals to absurdity.

We have entered the Age of Space. Already, the artificial *vamp*, like that in *Metropolis*, the film by Fritz Lang, is ready to replace the irresistible siren; the robot which kidnaps beautiful blondes in space operas is little by little replacing the ancient satyr, the persecutor of nymphs, and his numerous more simian descendants; flying saucers, coming to abduct earthlings, have replaced the roc-bird; little extra-terrestrials, more or less green, have taken the place of leprechauns and other gnomes; and the Bermuda Triangle all by itself has produced more victims than the Oceanic Trinity combined - the mermaid, the giant squid and the great sea-serpent. To compensate for the possible disappearance of maleficent creatures which menace us here on Earth, Science Fiction, through the voice of the Canadian writer Alfred Elton Van Vogt (1912 - 2000) has even proposed the existence of a *Fauna of space*, whose appetite for conquest will no doubt one day cause us all great distress.

Monsters are not about to disappear. Their survival is already assured.

24

Biblical leviathan, as drawn by Gustave Doré.

THE FOREVER UNFINISHED INVENTORY OF THE ANIMAL WORLD

'Every good encyclopaedia today should be above all an
encyclopaedia of our ignorance: we would discover there the
infinity of what is possible.'
(Louis Pauwels, *Planete*, 1962)

The world sometimes is astonished that there can still be unknown animals. Effectively, people imagine that we now know practically all of the fauna of our old globe, which has been criss-crossed in all directions over hundreds of years and more. Can we not confirm this by simply looking around us?

This is a misleading illusion. In reality, it seems that the inventory of the animal world is only just beginning. In the unanimous opinion of specialists, after having described a million well-defined species, we certainly do not know even a quarter of all of those which exist. This, at least, is what had been thought until recently. In these last years, as a result of the painstaking collection of certain small insects and complex calculations based on the subtle equilibrium of ecosystems, i.e., of natural communities of various forms of life, we have been led to conclude that the total number of living species, animals, vegetables and microbes taken all together, must amount to something between ten and fifty million.

Facing page: Elasmosurs by Zdevek Burian.

It is estimated today that, of the some two million species already described, plants and mushrooms comprise only 22% of this total (440,000 species) and the unicellular organisms scarcely 5% (10,000 species), with the large remaining portion of more than 70% (1.5 million species) belonging to the Animal Kingdom. In brief, we probably only know between a seventh and a thirty-fifth of the total of all existing animal species. It is not possible to even arrive at a simple order of magnitude. The scope of our uncertainty is, to put it quite simply, staggering.

The surest thing that one can definitively say is that a global census of all living things will never be completed. For the good reason that, according to the most recent estimates, the number of species which appear each day - some fifty on average - largely exceeds that of the species which we are destroying - scarcely about twenty per day on average. No better way could be found to emphasise that the number of animals remaining to be discovered is literally countless, and cannot ever be determined with exactitude.

WE MUST CLASSIFY BEFORE INVENTORYING

If it is impossible *a priori* to count hidden beings - those, among others, which are of interest to cryptozoology, i.e., those beings which are hiding, deliberately or not, and of which certain ones, if we are to believe the disbelievers, do not exist at all! - a study of past discoveries of hitherto unknown species, belonging to various zoological groups, could nonetheless enlighten us on two particular points: on the one hand, the relative numerical importance of each group within the framework of the Animal Kingdom and, on the other, the rhythm with which the descriptions of new forms have occurred over the centuries in each one of them. In any case, this should permit us to calculate, at least to a satisfactory approximation, the number of species which one can hope to discover each year in the immediate future. And that is what is of the greatest interest to cryptozoologists.

The inventory of the animal world has been going on since the most ancient times. However, a scientific system for doing this was only set up around the middle of the eighteenth century. In the course of successive editions of the *Systema naturae* of Carl von Linné, which drafted the taxonomy of living beings, in other words, the science of their classification and the rules which govern it, the Swedish naturalist in 1758 still distinguished only six classes, with sometimes indistinct outlines, within the Animal Kingdom: the *Quadrupeds,* which today are called the mammals, the *Birds,* the *Amphibians,* which then included the reptiles and even certain fish, the *Fish,* the *Insects,* which in fact included all of the articulate animals, the present arthropods and, finally, the *Worms,* the catch-all group to which were relegated pell-mell all 'soft' animals, that is, those lacking both a segmented carapace as well as the internal skeleton of the vertebrates.

Each of these classes was divided into a certain number of orders, and these orders in turn were divided into families, the families into genera, the genera into species, and the species into varieties. And, it was always possible to add to these major divisions all imaginable subdivisions, like the subphyla and the superclasses, the suborders and the superfamilies, the subgenera and the subspecies, which for the most part were only geographical races, to say nothing finally of certain complementary units which were created urgently to take account of awkward cases, such as the infraclass and the infraorder, and even the cohort, the legion and the tribe.

Let us forget these refinements. To take as an example the species which is the most familiar to us - our own - the zoologist will say that Modern Man belongs to the Animal Kingdom, to the phylum of the Chordata, to the subphylum of the Vertebrates, to the class of Mammals, to the order of the Primates, to the superfamily of the Hominoids, to the family of the Hominids, to the genus *Homo* and to the species *sapiens*. This means, whether we like it or not, that we are an animal, like the amoeba, the sponge, the jelly-fish, the earthworm, the mussel, the louse and the sea-urchin, a chordate, like an ascidian or any kind of fish, a vertebrate, like the shark, the toad, the viper, the vulture and the rat, a mammal, like the platypus, the kangaroo, the mole, the weasel, the vampire, the macaque, the armadillo, the pangolin, the rabbit, the mouse, the sperm whale, the hyena, the seal, the aardvark, the elephant, the daman, the manatee, the donkey and the ox, a primate, like the tarsier, the lemur and the baboon, a hominoid, like the australopithecus and the pithecanthropus, a *Homo*, like our unfortunate brother, Neanderthal man, but belonging more particularly, and in all modesty, to the species *sapiens*, i.e., 'wise'.

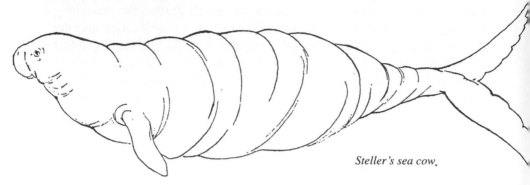

Steller's sea cow.

CERTAIN ANIMALS ARE DIFFICULT TO CLASSIFY

At the beginning of the nineteenth century, when the matter was being studied more closely, certain of the Linnaean classes were destined to be divided up little by little. Subjected to the painstaking analysis of such an attentive observer of animals as Jean-Baptiste de Monet, Knight of Lamarck, these classes soon amounted to twelve, in 1806, and then to fourteen, in 1809. The old class of the Worms which, in addition to true worms, included just about no matter what, even certain insects, had literally exploded. Outside of the Vertebrates, which are easily recognisable, there were then distinguished the Mollusca, truly soft creatures even under a shell, the Annelida, or segmented worms, the horde of Intestinal Worms, the ambiguous Cirripedia, namely, the stalked barnacles or goose-barnacles, and the acorn-barnacles or acorn-shells, considered today as crustaceans, the indisputable Crustacea, and then the Arachnida, the truly innumerable Insects, the Radiata, or organisms having radial symmetry, of which the starfish are the most characteristic, the Polyps, these usually marine animals, fixed in place like plants and, finally, the Infusoria, invisible to the naked eye.

On the other hand, however, Baron Cuvier, in his *Animal Kingdom* (1817), endeavoured to reduce these classes, which were proliferating alarmingly, to four quite distinct forms of organisation: the *Vertebrates*, the *Mollusca*, the *Articulata* and, at the bottom of the scale, the *Zoophyta* or 'animal plants', of equivocal behaviour. These fundamental types were going to give rise to what is called in our days the phyla, i.e., the large groups of which the filiation, or common origin, no longer appears in doubt in the eyes of the evolutionists.

With the progress in our knowledge of comparative anatomy and in the study of the processes of development, the four original branches of Cuvier, each one of which gathered within itself an endlessly growing number of classes, in turn underwent a gradual disintegration. Today, it is generally considered that there are at least about twenty phyla, which are subdivided into at least some seventy-five classes, not counting those which have totally disappeared in the course of the passage of time.

To give an idea of the richness and variety of the world of animals, it will suffice to recall the overall framework of the zoological system, by enumerating the various phyla which can be distinguished there.

Let us hasten to note that there have been an enormous variety of classifications of this sort; as many, in fact, as there are classifiers.

All, however, agree *grosso modo* on the essential points. They scarcely differ except in two aspects: *primo*, in the manner of associating or dissociating the numerous phyla of soft invertebrates which are not mollusks, all of those which were formerly classified among the worms and their aberrant cousins, the so-called vermidians, and, *secundo*, in the place which they provide for some disconcerting classes of arthropods. Among these ambiguous little grouplets there must be cited the Pycnogonida or Pantapoda, which are sorts of long-legged marine spiders, which, to be sure, have eight legs like all arachnids, but which strangely are provided with an aquatic respiration system, like the crustaceans. There are also the Tardigrada, microscopic beings with four pairs of clawed legs, which live in lichens and mosses and seem to curl up and die when these dry out, but which happily come to life again, even after years of apparent death, as soon as they are moistened. And then there are the Linguatulida or Pentastomida, little vermiform vampires which stealthily come as parasites into the internal organs of animals of all sorts. Finally, we must mentioned the Onychophora or Peripatidae, which resemble large caterpillars and which live in rotting wood or under stones in the shade of damp forests. The Swedish specialist, Stefen Bengtson, has described them in a most picturesque manner by saying that they appear to

31

have been born of lovemaking between a centipede and the Bibendum of the Michelin company.

Above all, it must be added, in concert with Pierre-Paul Grassé, long an outstanding leader of French zoology, and even of world zoology, thanks to his unequalled *Traité de Zoologie* (Treatise on Zoology), that, in the light of the continuing evolution of our knowledge, the zoological classification system is in perpetual change; 'One would be seriously mistaken to believe that the present systematics is definitive and emphasizes only natural lines and their respective relationships' (1969). Those who relish cryptozoology may rejoice; all classifications of animals, like all of their inventories and numerical estimates, are provisional and threatened with being overturned at any moment.

NO POINT IN LOOKING FOR
THE LITTLE ANIMAL HERE

Let laymen smitten only with a thirst for adventure, or curious people looking for marvels, and amateur naturalists as well, all be reassured. All those who have not travelled through the maze of zoological sciences in their multiplicity of byways, but who are passionately interested, all the same, in the exploration of the animal world, can forget straightaway the classifications of this kingdom; these forbidding and tedious lists of grotesque and barbaric names. From the point of view of cryptozoology, this science which aims to discover (in the proper sense of the word: 'discover' or 'uncover') hidden animals about which some information already exists, it is not necessary to be concerned with most of these existing branches. It is enough to know that they exist.

Just by eliminating all microscopic or truly minuscule animals, as well as those which are hiding in the organs of others or in the darkness of the ocean depths, we can already reduce the number of these phyla by half. For a still unknown being to be observed, without being captured at the same time, or for it to draw attention to itself, it goes without saying that it must be clearly visible and thus, above all, that it must be of appreciable size. Such a concept is nevertheless subjective and quite relative. It is thus proper to replace it instead with that of an abnormal size within a specified group.

Few people, other than professionals, would think of reporting the chance observation of an ordinary bird the size of a sparrow or of an ordinary-looking fish the size of a gudgeon. But, on the other hand, everyone would marvel at the sight of an ant as big as a shrew, or of a spider capable of covering a long-playing record, or of an earthworm comparable in thickness to a garden hose.

Indeed, apparently elusive animals which are much spoken of - those around which a legend has not taken long to build up - are those which are characterised above all by some truly unusual trait, a trait which is paradoxical, unexpected, striking, emotion-generating, and thus susceptible to mythification. It is not necessarily substantial bulk, but at least some impressive dimension remains essential.

So this is the reason why, in cryptozoology, we are only very rarely concerned with most of the phyla which have been entered into our classification. The fact is that the true giants - those which are measured in meters - are quite rare among the invertebrates. This exceptional stature contrasts with the general moderateness of size in these largely predominant groups.

Among the animals without vertebrae which are known to us, there is scarcely any one of them of a truly impressive size. At the most, one can cite a colossal jelly-fish, the Arctic cyanea (*Cyanea arctica*), the gelatinous dome of which can on occasion exceed 2.50 meters in diameter, and the stinging tentacles which extend over more than forty meters; a seemingly endless worm of the group of the marine nemertines (*Lineus longissimus*), not larger than one's thumb but measuring up to thirty meters in length; solitary parasitic worms of whales and sperm whales - gigantism calls for it - which reach lengths of 20 to 30 meters. Even a South African earthworm (*Microchaetes rappi*) can, in exceptional cases, exceed a length of 6 meters. Only the phylum of the Mollusca contains giants comparable to the most enormous vertebrates. One finds them among the cephalopods, along with ordinary octopuses and calamari, and their weight is measured in tons. The best-known is the supergiant squid (*Architeuthis*). Measurements on specimens of this genus have revealed bodies of around ten meters in size, excluding the fishing tentacles. Moreover, it must be recalled that, in the course of past geological eras, the largest invertebrates likewise were recruited among the cephalopods, in this case, the Ammonites, whose bodies are protected by a spiral shell. One of these, measuring nearly two

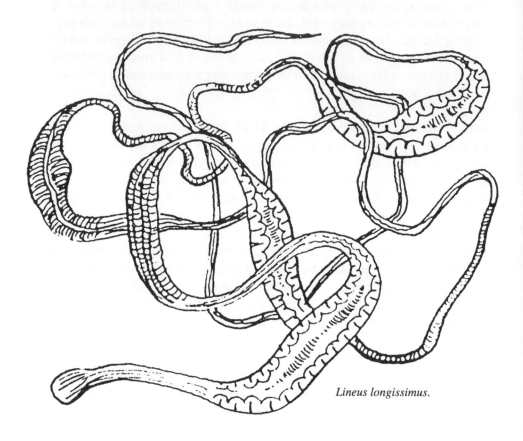

Lineus longissimus.

meters in diameter, was found in the cretaceous layers of Seppenrade, in Westphalia.

The Arthropods of very great size are still less spectacular. To our knowledge, no insect has ever reached a half-meter in size. Certain sea-scorpions of the Primary, or Paleozoic, exceeded only slightly the size of Man; their champion, a species of *Stylonurus*, measured at most three meters. In our days, the largest of the known crustaceans is the giant crab of Japan (*Macrocheira kaempferi*). Spread out flat, its size approaches four meters. This is far off the size of the largest whales, which do not fall far short of a length of 35 meters.

THE VERTEBRATES, OUR BROTHERS, OUR COUSINS, OUR RELATIVES

In brief, with rare exceptions, all animals of medium to large size are comprised - and, it is believed, have always been - in the last phylum of our classification, that of the Chordata, and again in a single one of its branches or subphyla, the last one.

These branches are, in fact, three in number: that of the Urochordata or Tunicata, the best known representative of which is an edible ascidian, the tasty *vioulet* or *biju* of the markets of Provence, celebrated in song by Gilbert Becaud; then, that of the Cephalochordata or Acrania (not having a skull), where there reigns on high the famous amphioxus (originally included in the genus *Branchiostoma*), a transparent marine dart, about a half-dozen centimeters in length, which is held to be close to the ancestral type of the vertebrates - in a certain way, one of our distant forebears; and, finally, that of the Vertebrata or Craniata (provided with a skull), which thus proves to be the principal target of cryptozoologists.

In short, except in the marine domain, where careful account must be taken of the possible existence of gigantic invertebrates, or at least those of an appreciable size - belonging in all likelihood to one or another of the four particular phyla (Cnidaria, Nemertina, Mollusca and Arthropoda) - the still unknown animals which one can seek to surround more and more closely, and to identify zoologically with the maximum of precision, are almost surely going to be found in one of the classes which compose the subphylum of the Vertebrata, namely:

a) the Agnatha, fish without jaws represented in our days by the lampreys and the hagfishes;

b) the Placodermi, comprising above all the armoured fish of the Devonian, but also, according to certain authors, the enigmatic chimaeras, abyssal rat-tailed fish which formerly were classified among the following;

c) the Chondrichthyes, or cartilaginous fish, which include the sharks and the skates;

d) the Osteichthyes, or bony fish, namely, all of the others which do not belong to any of the preceding classes;

35

e) the Amphibia, or creatures which lead a double life, primarily aquatic but also terrestrial;

f) the Reptilia, or crawling creatures;

g) the Birds, feathered winged creatures;

h) and the Mammals, or those having breasts, in general covered with fur when they are neither aquatic nor burrowing.

Ordinarily, each time that a 'new' animal is collected or captured, it is rather easy for the trained zoologist to determine in a brief examination - often with just a simple glance - to what phylum it belongs. Nevertheless, when the creature in question is one or another parasitic worm, or a marine vermidian, or an arthropod which is not clearly articulate, this can present certain difficulties, except in the case of certain rare experts. Even for these latter, a definitive classification can sometimes take years, and even then remain the subject of controversy. It is only if the unknown form belongs to the subphylum of the vertebrates that it is almost child's play to recognise the class in which it should properly be placed. Specialists in each of the fundamental groups - ichthyologists, batrachologists, herpetologists, ornithologists and mammalogists - will even be able to recognise straightaway the order in which it is to be recorded, and perhaps even its family. Determination of the genus, and above all of the species, for the most part requires a thoroughgoing study.

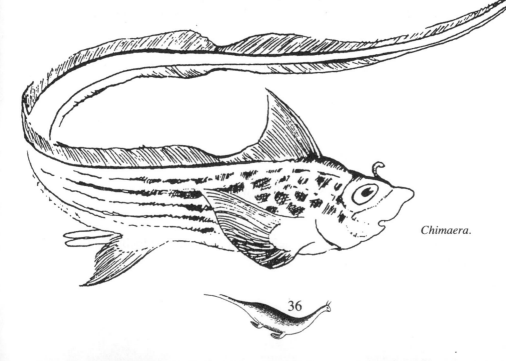

Chimaera.

WHERE THE DEVIL TO CLASSIFY THE UNCLASSIFIABLE?

It was soon recognised that creatures exist which do not correspond at all to the apparent definition which stems from their name. We have already mentioned worms which are not vermiform and articulates without articulation, but, among the vertebrates themselves there are fish which are going to promenade on land, amphibians which are solely aquatic or solely terrestrial, reptiles which, far from crawling, prefer to gallop, leap, glide or fly, birds, on the other hand, which are incapable of flight and, finally, mammals deprived of true nipples. Nature abounds with a perverse imagination.

Nevertheless, since all of these deviants possess the majority of the other traits of their group, it is generally easy to recognise their true nature. But, does it not occur that one truly does not know where to classify an unknown animal, for the good reason that it does not fit within any of the known and recognised groups?

This can in fact happen and, moreover, it has occurred frequently in the past. It even continues to happen now and then.

There was a time, not so long ago, when it was not even known to what kingdom certain living forms should be assigned. Until the eighteenth century, coral was considered to be a petrified marine tree, and it was believed that it belonged both to the Animal Kingdom and to the Vegetable Kingdom. This dual nature was also attributed to those creatures which were, significantly, termed the zoophytes, namely, the sponges and the sea anemones, of which, according to Pierre Belon (1555), it was not really known 'if they are plants or animals'.

To be sure, a profound study effectively dissipated the ambiguities, which were, after all, only superficial. And, over the centuries, when a new animal of our times burst upon the zoological scene, the means were always found to lodge it in one or another of the known phyla and even in a known class, for the frontiers of these vast groups long remained poorly defined. The existence of catchall categories, like that of the worms, the vermidians, the articulates or the zoophytes, was in this respect quite providential. Nonetheless, even when the diagnosis of the fundamental categories was

37

carried out with rigour, problems of classification still arose from time to time, and it was occasionally necessary to create new groups, for the classification of the unclassifiable. This has still happened in quite recent years.

To be sure, it is above all the exhumation of fossil forms having no precedent, which generally requires the creation of more or less vast new categories. On the other hand, when a still unknown living form is discovered, it is sometimes found that it can be placed within an already established group, but a group whose representatives had been known up until then only by their fossil remains. In such cases people tend to cry out 'living fossil', a paradoxical expression invented by Darwin and which has been much misused, since, to tell the truth, it really means nothing. We shall come back to this in more detail in the next chapter.

Whether the original forms are recorded among these survivors from the past, in other words, among those which have disappeared forever or simply among those which have succeeded in preserving their anonymity, the frequency of such discoveries logically should decrease, as a consequence of the constant progress in our exploration of the Earth and of its geological strata, the number of species remaining to be discovered thus diminishing little by little. But, we already know that the latter are so numerous that we do not need to be apprehensive about the richness of the harvests yet to come. This being said, the abundance of new units of classification, created to receive the newcomers, will necessarily diminish over the course of the coming years.

In our days, the description of new species is a routine affair, and that of new genera still remains quite commonplace. It is much more rare that one is obliged to establish a new and quite distinct family in order to receive an unknown. As to innovations at the level of higher categories - orders or classes - these are, as one may guess, quite exceptional.

However, they are not quite as exceptional as one would imagine. As extraordinary as it may seem, in our times it still occurs fairly often that it is necessary to revise rather considerably the classification of animals, in order to enter entirely new and completely unexpected forms. Thus, the most astonishing event really almost unthinkable would evidently be the discovery in our days of a whole unknown phylum of the Animal Kingdom.

Nevertheless, this has already happened three times since the beginning of the 20th century, the last time being less than ten years ago, and it is possible that this could happen again tomorrow.

To judge from this summary of the most recent upsets in zoological classification, there are still promising days ahead for researchers with open and inquisitive minds.

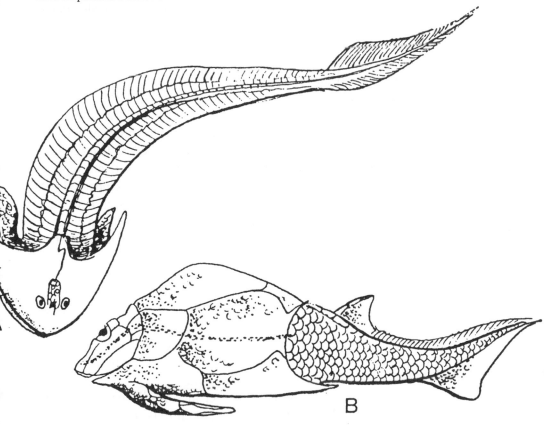

Armoured fishes of the Devonian period: A. Cephalaspis; B. Pterichthyodes.

39

THE PROGRESS OF OUR EXPLORATION OF THE ANIMAL WORLD

If, from the balance sheet in question, we seek out some practical information, we note first of all that the most striking innovations of the twentieth century have come from the sea and, moreover, that they involve almost exclusively invertebrates and, further, that these are almost all of small size, gathered in general in the great depths of the sea, but also in the no less dark waters of caverns.

If, among the terrestrial vertebrates and even among vertebrates in general, we are to cite major events comparable to those which in our time have revolutionised the vast world of the invertebrates, we must revert back to the nineteenth century. It was in those days, for example, that there was discovered in New Zealand the sphenodon or hatteria, the little spiny lizard which the Maoris called *tuatara*, and which was judged to be a late survivor of the order of the Rhynchocephalia, known only from the fossil state: reptiles older than the prestigious dinosaurs and presumed to have disappeared 135 millions of years ago. It was also at this time that it had been necessary to invent a whole new subclass of mammals; the Monotremata, in order to classify the platypus and the echidnas of the Australian region, which had the indecency to lay eggs, like reptiles or birds.

However, this does not at all mean that comparable revolutions cannot still occur in our days. After all, one of the most stupefying zoological discoveries of modern times remains that of the *coelacanth*, which took place on 25 December 1938, a Christmas gift which no zoologist in the world ever would have dreamed of finding in front of his fireplace.

Certainly, it would appear that we have little chance of still encountering, at least in known lands, a large new vertebrate, of entirely original structure, and thus unknown even in fossil form. But is that not asking the impossible? The vertebrates constitute only one subdivision of one of the some twenty-five phyla which make up the Animal Kingdom. What should provide comfort and encouragement to those seeking surprising discoveries is that one can still discover hitherto unknown families of terrestrial vertebrates, both among the Mammals as well as among the Reptiles and the Batrachians. And, it must be stressed above all that the discovery of a simple new animal species, provided that it offers something unusual or really original, can not only make page one of the newspapers, which is of negligible value from the scientific point of view, but can also clarify considerably certain zoological problems.

So this is why it is important to seek to determine mathematically, to an acceptable approximation, the number of species which one can hope to discover in the decades to come, by starting from the number of species catalogued little by little in the course of the past centuries.

Throughout the history of zoology, several authors have tried to carry out this work of extrapolation. The results of their calculations are sometimes quite baffling, for they bring out clearly some brutal irregularities, often totally incomprehensible, as well as downright absurdities. For example, in certain groups, there appear to have been many fewer species known at a given point in time than was the case in an earlier period.

There are a number of factors which help to understand these inconsistencies. For one, the same species have often been described several times under different names, which has caused many of them to fall into synonymy. For another, certain inventories deal with all known forms, namely, the subspecies as well as the species. Now, for mammals and birds, for example, the number of subspecies, or geographic races, is about three times as great on average as the number of species. And, finally, a great number of species of former times turned out to be sometimes a representative of the opposite sex with respect to the type-specimen, sometimes a different phase of development, sometimes an accidental hybrid, and again sometimes an individual variation, stemming either from teratology or from genetics (a colour mutation, for example). The elimination of all of these pseudo-species readily explains the apparent decreases.

And then, two opposing tendencies have long divided the classifiers. Depending upon their approach, they have been termed either 'splitters' (those who split or divide up) or 'lumpers' (those who lump together). Today, it is rather the latter who dominate, for which one may be duly grateful. Still, excesses should be avoided. Here as in all things, truth is to be found in the golden mean.

A NUMERICAL ASSESSMENT OF OUR HOPES AND EXPECTATIONS

I, Bernard Heuvelmans (1983), am responsible for the most recent research on the rate - variable in any case - at which there has occurred and continues to occur the progressive discovery of the animal species populating the planet. This work was based on the comparative juxtaposition of the various inventories of the past and the hypothetical correction of each of these in the light of their context and the data which they contain. By means of graphical representation, the study set out to measure, as a function of time, the enrichment in species of the various groups of the Animal Kingdom, from Linnaeus to our days.

To be sure, the old classifications were not as detailed or as judicious as those of our time, so it was these former which it was necessary to take as a base of reference for the sake of unity, both in the tables and in the figures. There were thus recreated antique depositories like that of the *Vermes* (worms and vermidians), and there were likewise associated again, in the heterogenous group of the *Protochordata*, all of the chordates other than the vertebrates, and the *Hemichordata*. Thus, there were assembled together as formerly, beings as little related to one another as the Mollusca and the Brachiopoda.

The various curves obtained (by a judicious smoothing out of the zigzags resulting from the graphical representation of the raw numerical data) represented definitively, and as faithfully as possible, the probable increase in 'good' species in our zoological catalogues in the course of the last two centuries. On the one hand, they covered the numerous old groups of the Animal Kingdom and, on the other, the various classes of vertebrates as well, which are of particular interest from the cryptozoological point of view.

The extension of curves beginning in the past made it possible to obtain, by extrapolation, fairly precise information on the situation in the immediate future. It was also possible to make predictions for the longer future, even though more uncertain, and notably for the year 2000. These predictions plainly become all the more risky as the slope of the curve becomes steeper, in other words, as the number of species described annually becomes greater. Consequently, it is the data relative to the Mollusca and, especially, to the Insects that are the most open to question.

42

This being said, even though since the establishment of the curves in question there have become available a number of additional general inventories, as well as several rigorous counts in connection with the Vertebrata, these graphical representations on the whole remain valid and, moreover, have been confirmed by findings on many specific points.

The translation back to numerical data of all of these curves enables us to establish a significant picture of our progressive exploration of terrestrial fauna. In this way it is found that the number of known animal species increased by close to 350,000 units in the course of the second half of the nineteenth century and by about 600,000 units during the first half of the twentieth century, and that it will increase further by more than 250,000 units: at least (from a conservative point of view) throughout the second half of the twentieth century. The inventory of the animal world thus increased annually, on average, by about 7000 new species between 1850 and 1900 and by some 12,000 new species between 1900 and 1950, and this average increase will still be at least 5000 per year between 1950 and 2000. In this annual harvest it is, of course, the insects which will take the lion's share (or the ant-lion's share, perhaps one should say!): around 4750 annually during the second half of the nineteenth century, more than 9000 during the first half of the twentieth century, and 4000 at the very least during the second half.

In short, it was in the course of the first half of the twentieth century that the rate of discovery of still unknown species reached its peak for the majority of the phyla of the Animal Kingdom. The rhythm of their descriptions - their tempo - then slowed in a more or less abrupt fashion, depending on the phyla under consideration. This slowing down is all the more appreciable in our days, in view of the important number of discoveries which occurred in the more favourable period of the first half of the century. But again, is this not logical? The more rapidly the inventory has grown, the more the stock remaining to be inventoried becomes scarcer earlier, and thus the rarer become new descriptions.

That which has happened for the various classes of Vertebrata should above all be the focus of attention of the cryptozoologist. Around 1975, there were still being discovered each year 118 species of fish, 18 species of reptiles, about ten species of amphibians and equally as many mammals, and only 3 or 4 species of birds. But here as well, the tempo of these discoveries was slowing differently, according to the classes. This can be seen in the most striking manner from a graphical representation of this phenomenon.

43

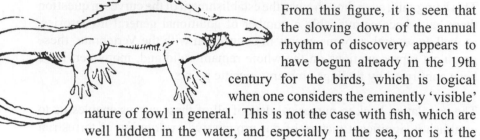

From this figure, it is seen that the slowing down of the annual rhythm of discovery appears to have begun already in the 19th century for the birds, which is logical when one considers the eminently 'visible' nature of fowl in general. This is not the case with fish, which are well hidden in the water, and especially in the sea, nor is it the case with mammals, which are much more elusive than one would imagine, doubtless because of the considerable development of their intelligence: for these two classes, a true decline began to be noted only around 1925. As for amphibians and reptiles, which are particularly wary, the rate of their discovery has scarcely slowed at all in the course of the years.

On the basis of this graphical representation one would be led to believe that, toward the year 2000, there will still be discovered each year some 90 species of fish (which is a relatively small number, given the numerical importance of the species in this group), 7 species of amphibians and 3 or 4 species of mammals. It is also to be thought that at that time the census of bird species will have been completed definitively.

This being said, my investigation having been based in places on certain controversial premises, concerning which there is no consensus at all, the numerical conclusions which are drawn from it are tarnished with a certain lack of precision. But this is true in the sense of optimism as well as in that of pessimism. Thus, the precise counting of the number of species of amphibians showed that, in 1985, this exceeded by about a thousand, that is, by about a third, the highest number estimated. Thus, in this class the rhythm of discovery of unknown species has not even begun to decline.

At all events, the undertaking aimed essentially, by my own admission, to define relative situations and to give orders of magnitude. My conclusions simply permitted the greatest hope of discovery for the future and thus justified fully the utility and even the necessity of cryptozoological research, of which I am the leading advocate.

THE IMPRUDENT PREDICTION OF BARON CUVIER

Without doubt the objection will be raised that even the quasi-certainty of discovering each year, in the course of the decades to come, some 5,000 species of unknown animals, including 150 vertebrates, does not warrant the expectation that there will be among them forms of respectable size, among which are to be found the majority of those which are the target of cryptozoology.

But, a celebrated precedent is going to show that such a pessimistic objection does not stand up. It is not a matter here of vague estimates or of far-fetched conjectures, but of verified and indisputable facts.

In 1812, Baron Georges Cuvier, in the introduction to his *Recherches sur les ossemens fossiles*, declared unequivocally: 'There is little hope of discovering new species of large quadrupeds'.

For him, the contemporary fauna were at that moment all well known. All animals, even those of the newly explored regions of the Americas and Australia, had been recorded and described. Naturalists had thus henceforth to concentrate their attention on vanished animals, for which, in fact, an enormous quantity of fossilised remains were beginning to be dug up.

What exactly did the one who is justly held to be the Father of Paleontology mean by 'large quadrupeds'?

At the time, even though lizards and crocodiles, frogs and newts were also provided with four paws, the name 'quadruped' was used exclusively to designate what we call today 'mammals'. As to what Cuvier considered as mammals of a fairly appreciable size, one can judge by the list, which he went on to quote, of the previously unknown animals discovered in the new lands: 'the puma, the jaguar, the tapir, the capybara, the llama, the vicuna, the sloths, the armadillos, the opossums, all of the capuchins' in southern America, and 'the various kangaroos, the wombats, the dasyures, the bandicoots, the flying phalangers, the platypus, the echidnas' in Australia and in

the adjacent islands, i.e., in all of Australasia. Among all of these 'large quadrupeds' there are several, like the dasyures and the platypus, which do not weigh even so much as two kilos in the adult state, and certain ones, like the flying phalangers, do not even reach one kilo.

Let us now strive to replace here general descriptions of size with terms which are more rigorous and more scientific.

The weights of adult mammals range between two extreme values: about two grams for the bumblebee bat (*Craseonycteris thonglongyai*) and the pygmy shrew or Etruscan pachyura (*Suncus etruscus*), and about a hundred tons for the blue whale (*Balaenoptera musculus*). In the following table, the different sizes (expressed in weight) are set forth in geometric progression with a ratio of ten, which permits an objective classification, by order of magnitude, of the various categories of mammals.

1,000 tons

} GIANT (more than 50 tons)

100 tons

10 tons

} LARGE (between 500 kilos and 50 tons)

1 ton

100 kilos

} MEDIUM (between 5 and 500 kilos)

10 kilos

1 kilo

} SMALL (between 50 grams and 5 kilos

100 grams

10 grams

} TINY (less than 50 grams)

1 gram

In a recapitulation of the 'large quadrupeds' described since the imprudent statement of Cuvier, we shall, to be sure, take no account whatever of mammals of small size, nor *a fortiori* of the tiny ones. And, we shall also take no account of aquatic mammals, and above all the marine mammals (cetaceans, sireneans and pinnipeds), which nevertheless include the only giants, but which Cuvier chose to ignore, although not entirely unwisely. We shall pass in review only the terrestrial mammals having at least the size of a domestic cat or, to express the matter with more precision, those in which adult specimens approach at least a weight of 5 kilos, even if only exceptionally.

Unknown Antartic cetacean seen by Edward Wilson in 1902.

47

A PESSIMISM UNCEASINGLY BEING CONTRADICTED

This panorama is dreadfully and overwhelmingly grievous for one who remains one of the greatest names in zoology. Before his death in 1832, Cuvier had already been contradicted by events ninety times. Nevertheless, in the sixth, corrected edition of his famous introduction (1830), published henceforth under the title of *Discours sur les Révolutions de la Surface du Globe (Treatise on the Revolutions of the Surface of the Globe)*, he still - with his habitual lack of honesty - recognised only one defeat: 'The single major exception which one can bring up to me is the tapir of Malacca, recently sent back from the Indies by two young naturalists among my students, Messrs Duvaucel and Diard, and which in effect constitutes one of the finest discoveries with which natural history has been enriched in recent times'.

The truth is that, from 1814 to 1847 - during a third of a century - *not one* single year passed without at least one species of 'large quadruped' being described. In certain years there had even been some ten of them described. Altogether: 171. And, among these, there were some truly very large ones: several rhinoceroses, two horses, buffaloes, stags, enormous antelopes, a hippopotamus, wild boars, bears and anthropoid apes.

From 1849 to 1912 or, in other words, in the course of the following two-thirds of the century, the rhythm of discovery of terrestrial mammals of medium to large size - 125 in all - slowed down only by half. In essence, in the course of the first century which followed the peremptory declaration of Cuvier, there were discovered 296 'new species of large quadrupeds' - to use his own words - or a total of nearly three hundred.

Finally, from 1918 to 1990 - a period which may be termed 'our times', i.e., the two-thirds of a century which followed the First World War - not less than 38 species of mammals of an appreciable size were still described.

In brief, 334 species incapable of passing unnoticed - an average of two per year - made themselves known after Cuvier had declared that there was little hope of still discovering any such. And, it must be recalled that, at that time, the great paleontologist of Montbeliard himself knew only a total of 386 species of mammals of all sizes. Since then, there have been discovered almost as many as he himself had termed 'large'.

Such an undoing of his predictions did not discourage the pessimists in the least or, more exactly, it did not discourage those who appear to be frightened by anything new because it disturbs their intellectual comfort. Certain zoologists of a conservative turn of mind, in fact, persist obstinately in repeating the substance of the assertion of Cuvier, even though this has been shown to be gross foolishness.

Witness, for example, what was said in 1934 by Dr Charles Anderson, of the Australian Museum: 'Although there are many kinds of land animals yet to be discovered and described, they are mostly of small size, and it is safe to say that there is no mammal, bird, or reptile of large dimensions and unusual structure which is entirely unknown, and which would not fall naturally into some well recognized group'.

The falsity here is easily exposed. The still hidden animals which one can no longer hope to discover in our days are not only just those of large size, but they must also be of an unusual structure and even clearly unclassifiable. In short, the statement amounts to an assertion that there remains little chance of discovering new phyla, like those of the Pogonophora, the Gnathostomi or the Loricata, but which would not be composed, as they are, of tiny marine animalcules, but, rather, of large land animals. There, assuredly, is something which 'it is safe to say'. But what cryptozoologist has ever claimed to pursue such an objective? Even those who are striving to track down presumed dinosaurs in the Congo are, ultimately, seeking only animals which are known in the fossil state, which have a normal anatomy, and which fit naturally into a well recognised group.

THE INCONSISTENCIES OF PROFESSOR SIMPSON

In the face of the perfect legitimacy of the hopes and expectations of cryptozoology, which are based on irrefutable lessons of the past, it was clearly necessary, in order to combat these, to resort to falsehoods and disinformation, which in fact has been done with enthusiasm. In order to make cryptozoologists look ridiculous, as has always been done with all innovators, it was necessary to attribute to them intentions or aims which they never had, like being interested exclusively in 'monsters' (a term which is equivocal *par excellence* and virtually undefinable) or in terrifying beasts which have emerged from Prehistory. Some have gone so far as to stack the deck by shamelessly falsifying numerical data.

49

Of particular significance in this regard is the very latest article by the great American paleontologist Georges Gaylord Simpson, devoted to mammals from the cryptozoological point of view (1984). Perhaps apprehensive, as was Cuvier formerly, at the thought of seeing cryptozoological discoveries in the domain of present fauna undermine the paleontological edifice which he had so laboriously built up in the course of a long and brilliant career, Professor Simpson gave himself over to a well-ordered attack on the new discipline, which had recently been made official through the establishment of the International Society of Cryptozoology.

Other than the fact that this article abounds with errors, grave omissions and whimsical interpretations, which may perhaps be attributed to the advanced age and declining state of health of the author (who was to die shortly after having published it), some absolutely flabbergasting assertions are to be found in it. In particular, Simpson writes here that, in the course of the past fifty years: 'there has been no definite and objective discovery of any living taxa that were previously unknown or hidden in the cryptozoological sense'. If one goes back to the definition which the paleontologist himself gave for the *taxon* in 1965 ('a group of organisms officially recognised and named in a technical classification'), it must be accepted that every species designated by a valid Linnean binomial is indeed a taxon. Did Simpson, although an eminent classifier of mammals, really mean that *no* species of this class had been discovered since 1934? In any event, according to the catalogue of Honacki, Kinnman and Koeppl (1982), to which he himself refers, there have been 461 descriptions from 1935 to 1981. Furthermore, he cannot have taken into consideration only those taxa of generic rank, since he recognises in his article that 24 new genera have been discerned since 1940. And, he cannot even be alluding to the absence of any discovery of taxa at the level of families, since he admits that two of them have had to be created within the framework of the class of mammals in 1938 and in 1974, i.e., the Seleviniinae and the Craseonycteridae respectively.

Certain oversights by G.G. Simpson border on a pure and simple travesty of the truth: for example, when he says: 'No large living mammal, not to mention dinosaurs, has been discovered in Africa since 1901'. This is to put down and denigrate the descriptions of the hylochoerus, or giant forest hog in 1904, of the mountain nyala in 1910, and of the pygmy chimpanzee in 1924, which, although of a somewhat smaller size than the others, is nonetheless a great anthropoid ape.

50

The blindness of Professor Simpson reaches its peak when he declares: 'there is absolutely no objective, autoptical [that which one can see with one's own eyes] evidence for any new species of primates since 1907 at the latest'. Again, according to the catalogue of 1982, which he makes use of as a reference work, there were not less than 8 new species of primates described between 1908 and 1981, of which 6 were of large size. Their names will be found in our panoramic summary, which likewise is based principally on the list of Honacki, Kinnman and Koeppl.

DISINFORMATION AGAINST CRYPTOZOOLOGY

Emboldened by the immense prestige of Georges Gaylord Simpson, further enhanced by his recent death, a few researchers of lesser stature hastened, to be sure, to follow in his footsteps in order to go to war against cryptozoology, but without remarking the incoherence of the propositions of the great scientist, nor taking the trouble to verify his allegations. Among these must above all be cited Jared M. Diamond, a physiologist at the School of Medicine of the University of California at Los Angeles.

True, Dr Diamond recognises that 'new birds and mammals turn up every year and occasionally even a species thought extinct is rediscovered'. But, according to him, this 'does not satisfy the cryptozoologists' desire for monsters'. And, he goes so far as to claim that 'in fact, no new species of a large terrestrial mammal has turned up since the kouprey [...], discovered in Indochina in 1937'. The notion of size being entirely relative, this assertion could in a strict sense be defended, but one would have to be of singular bad faith not to consider as being of large size creatures such as the Himalayan blue goat, described in 1963, and the Paraguayan tagua from the Pleistocene, the largest known peccary, rediscovered living in 1975.

Further, the simple lack of honesty of Jared Diamond shows through in the recapitulatory table of supposedly mediocre zoological discoveries of the twentieth century. This table deals solely with 'new genera of mammals described from 1900 to 1983 and new species of living birds described from 1934 to 1975'.

One cannot compare that which is not comparable. Why not put in the same broad chronological period all existing data, which are in any event easy to collect? And, above all, why were the results obtained cited for species as

51

far as birds were concerned but *genera* as far as mammals were concerned? The answer is evident. Since it is a matter of stressing the paucity of recent innovations, the American biologist abstained from bringing in for the birds that period when the harvest of species was still very abundant. And, it was for the same reason that he deliberately confined himself to the number of *genera* of new mammals (126), much less impressive, to be sure, than the number of new species (about 1600).

One can, in fact, measure numerically, i.e., scientifically, the degree of insincerity of the author in question, in the light of one of his own assertions: 'As with mammals, the discovery rate plummeted in the first decades of the twentieth century, but has remained steady, at around three species per year since 1941'. If one trusts the meticulous catalogue of Honacki, Kinnman and Koeppl, as one should, it will be noted that from 1942 to 1981, or over 40 years, there were described 387 new species of mammals, in essence, an annual average of 9.6, which rather approaches *ten species per year.*

Consult, then, the unceasing discovery of large land mammals, and let the facts and the figures speak for themselves, which should enable anyone to re-establish the truth and to see where are to be found the competent, serious and honest men of science.

To deal both with Cuvier and Simpson, as well as with their onlookers, it is well at the end not to limit oneself to recalling the discovery of new mammals. Not less surprising have sometimes been the descriptions, over the course of decades, of new amphibians and new reptiles and, even more stunning yet, of new birds, namely, of flying animals which often are brilliantly coloured, in brief, the most showy in the world.

Crocodile belonging to the group of Thalattosuchia.

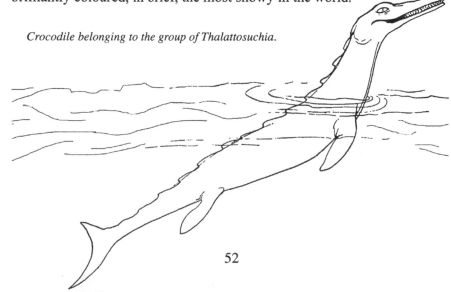

SURVIVORS FROM THE PAST

'In the mastodon, in mammon, in the paleontherium; in the giant dinotherium, in the ichthyosaur, in the pterodactyl, is there not all of the incoherence of a dream? Matter in the form of a nightmare, it is Behemoth. Chaos produces a beast, it is Leviathan. To deny these beings is difficult. The bones of these dreams are in our museums.'

(Victor Hugo, *Promontorium somnii*, 1863)

'Living fossil' is not, as one would be inclined to believe, an insult born of the generational conflict, one of the various synonyms for epithets such as 'doddering' or 'about to kick the bucket'. To the contrary, it is a cry of wonder before the very acme of paradox: an anachronism in flesh and blood. It is an expression which has – alas! - been used, misused and abused.

However, it would be in order to define it rigorously, for the question often arises in cryptozoology of the resurrection of fossils or, more exactly, of the *possible* survival of animals held unanimously to be extinct. Now, what could merit the term of 'living fossil' more than a being bursting forth, full of life, from the mists of time?

It is a fact that certain beasts which are still ignored by zoologists appear to be well known to paleontologists, even though in the state of petrified

Above: Basking Shark

remains. But, the former as well as the latter often refuse energetically to admit them to the circle of recent fauna. The reasons for their veto are all the more obscure, in view of the fact that our planet swarms with 'living fossils'. The more one studies with care those beings to which this title could be applied, the more of them are discovered, and the more they sink into the commonplace, to the point that one ends up by wondering if *all* living things are not in some way 'living fossils'.

WHAT IS A 'LIVING FOSSIL'?

To have a clear conscience, one must discover first of all the precise sense that the person who created this expression intended it to have. This is none other than Darwin, the father of the theory of evolution, and here is the passage from *The Origin of Species by Means of Natural Selection*, where he set it forth in 1859: 'And it is in fresh-water basins that we find seven genera of Ganoid fishes, remnants of a once preponderant order; and in fresh water we find some of the most anomalous forms now known in the world as the Ornithorhynchus and Lepidosiren which, like fossils, connect to a certain extent orders at present widely sundered in the natural scale. These anomalous forms may be called *living fossils*; they have endured to the present day, from having inhabited a confined area, and from having been exposed to less varied, and therefore less severe, competition.'

Thus, we see that, in the spirit of Darwin, the designation 'living fossil' should be reserved for rare forms surviving only in little 'lost worlds', isolated as though by miracle, and each of which at the same time represents an intermediate stage between two distinct lines. However, if one can believe the enormous body of literature devoted to 'living fossils' of all sorts, this idea must have undergone astonishing evolution, since it has been considered that it could legitimately be applied to the entire fauna of islands as extensive as Madagascar, New Guinea or New Zealand, and even to an entire continent, like Australia, as well as to entire classes (Crinoidae or sea lilies, Monoplacophora or the present pilinas, Xiphosura or king crabs, Onychophora or velvet worms) and even to entire phyla (Brachiopoda and Pogonophora). These vast groups all include numerous species and, with the exception of the velvet worms, in no way constitute transition forms.

Thus, it is hardly astonishing that, in the master-work devoted to this question, *Formes Primitives Vivantes* (*Primitive Living Forms*) (1970),

Claude Delamare-Deboutteville, of Paris, and Lazare Botoşanéanu, of Bucharest, ended up by giving the following simple definition of 'living fossil': 'a being which is still living in nature at the present time and which represents a former state in the cycle of evolution'. It is even less astonishing that these authors prudently added: 'At first, this definition may seem somewhat broad as, in principle, this designation is reserved for animals from a very distant past'.

The fact is that the definition in question can apply practically to any living species, since all have at least, certain archaic and *relatively* primitive characteristics which are more developed and more specialised in other forms. As for setting a limit on what is 'a very distant past', this, to be sure, will be arbitrary in any event; just as all definitions which have been given for the notion of 'living fossils' are conventional, as we shall see.

THE COMMONPLACE NATURE OF 'LIVING FOSSILS'

Everyone believes at first that he understands what is a 'living fossil', but this juxtaposition of two apparently contradictory terms is, in truth, difficult to define. First of all, the contradiction is really only apparent. In no way is it a question of an antinomic formula like 'a door must be open or closed'. One cannot say: 'a species must be living or fossil'. If it is true that all fossil forms have not persisted up until our days, each living form has nonetheless had ancestors which can formerly have had the exceptional privilege of being fossilised.

To claim that the expression 'living fossil' is applicable to beings which are still living, whereas they should *normally* (or *logically*?) have disappeared, is perhaps legitimate but still does not resolve the question. What damnation would a species carry within itself for having been condemned in this way to an unavoidable extinction?

In the eyes of the public at large, 'living fossils' are creatures of an extraordinary appearance, extremely rare, confined to almost inaccessible regions, and dating from a past age.

Of extraordinary appearance? Many people wash each day with the skeleton of a 'living fossil', doubtless several hundreds of millions of years old, and they have never found it anything other than completely ordinary. The fact is that sponges - the Porifera - are very ancient animals and have evolved almost not at all since the most ancient times. At the beginning of the Primary Era, in the Cambrian, there already existed sponges which are quite similar to those of our bathrooms. And that was more than 500 millions of years ago.

We also eat them, in months with an 'r' (I am talking of oysters, or course) - *à la marinière, à la provençale* or *à la belge* (with French fries!), according to our tastes - 'living fossils' which go back some 200 millions of years, and this does not astonish us in the least. The *mussel*, the modest mussel, in fact dates back to the Trias. The coquille Saint-Jacques, or *scallop*, is a little less ancient - it comes to us from the Lias or lower Jurassic - and the aristocratic *oyster* dates from the more recent Jurassic. As for the present *sea-urchins*, certain ones are as old as these various mollusks. In any case, here is a choice of 'living fossils' which make excellent *hors d'oeuvres*!

Even our most familiar domestic animals could, strictly speaking, be considered as 'living fossils'; even though relatively young, *dogs* go back to the upper Miocene and *cats* to the lower Miocene, i.e., to some twenty and thirty millions of years ago respectively, which in any case is not negligible.

Extremely rare, 'living fossils'? One which most has the appearance of what they are is without question the so-called *king crab*. With its lance like that of a knight of the Middle Ages and its rounded armour concealing monstrous spiders' legs, it looks like something which escaped from the Hell of Hieronymus Bosch. And, manifestly, it served as the model for the dragon of the city of Mons, the lumeçon.

Because it lives in sea water, the British call it *king crab* and some Americans *horseshoe crab*, but in reality it is more scorpion than crustacean. In the course of its development it passes through a 'trilobite' stage, thus betraying its affinities with a group which flourished during more than 200 millions of years, through the first half of the immense Primary Era. The genus *Limulus* itself existed already at the beginning of the Secondary. Today, five different species of it are to be found, which, moreover, are distributed

over three genera. Four species frequent the coasts of Japan and those of the Sunda Isles, and a fifth lives on the Atlantic shores of North America, from Maine to Yucatan. Despite its 200 millions of years of age, the king crab is so abundant that its pulverised carcass is used as fertiliser. By way of indication in this respect, a half-million specimens were collected for this purpose at Bowers, in Delaware, in 1927.

WE ARE ALL 'LIVING FOSSILS'

It is not necessary to run to the end of the world, or to dive to the bottom of the sea, to find a 'living fossil' as old as the famous coelacanth (which, we should mention in parentheses, lives at depths of scarcely a hundred to two hundred meters). Even in your cellar, hundreds of 'living fossils' from the Devonian and the Carboniferous patiently weave their webs, and are the despair of housewives. The tarantulas of the ancient red sandstone deposits of the County of Aberdeen already very much resemble our present spiders; and, in the coal deposits of England, Bohemia, Silesia and Illinois, there have been found remains of Arachnida of recent type. And if, in your library, you have very old books, or an old herbarium, or even a collection of insects, and if sometimes you see running through them the animal which is called a 'book scorpion', then you should know that these pseudo-scorpions, otherwise known as *Cheliferinea* already had close relatives in the Silurian, more than 400 millions of years ago.

True scorpions, quite similar to those of today, lived on our lands in the Carboniferous, 300 millions of years ago. In that epoch there were already swarms of cockroaches like those which today are constantly striving - sometimes victoriously, alas! - to take over our homes and our ships. Also in those times, numerous dragon-flies soared through the air, while grasshoppers hopped about on the land.

Later on, beginning with the Jurassic, many of our familiar insects were present: the may-bugs, the weevils and the beetles. The cicadas were already preparing to delight men - who were going to appear only 150 millions of years later - or else to deafen them. The flies did very well without our presence. And, the honey-bees worked industriously for their own strictly personal needs. Who among us would think of calling these little animals by the prestigious term of 'living fossils', although they are just as entitled to it as are their larger contemporaries, fish, amphibians and reptiles?

Eurypterida or sea-scorpions by Zdenek Burian.

Confined to a very limited little region, the 'living fossils'? There is at least one continent which is said to be inhabited only by such relics: this is Australia, the kingdom of the Monotremata and the Marsupiata. All of the other terrestrial mammals there have been imported by man.

Dating from an age long past? Yes, beyond any possible discussion, here is a trait which characterises all 'living fossils', without exception. Still, it must be recalled that *our* age, the Quaternary Era, which covers the Holocene and Pleistocene periods, began only 3 million years ago. And, at that time, the various forms of life had already long since been born. The fact must be accepted - there is really nothing new under the sun! - that all of the phyla which are known to us had become differentiated in the Paleozoic or Primary Era, and all of the classes which presently are known were established in the course of the Mesozoic or Secondary Era. The Cenozoic or Tertiary Era witnessed only a simple diversification of existing classes.

Delamare-Deboutteville and Botošanéanu laid careful emphasis on it: 'The world is aging very slowly. Almost all of the groups which surround us are, in a certain manner, living fossils on the scale of the group. According to general zoology, we are still living in the middle of the Devonian. We know precisely that the major groups of our time already existed at the onset of this epoch, and already had the fundamental characteristics of the present forms.'

To understand well the whole of the evolution of the various groups of living beings, it is essential to grasp such truths to the fullest. The living world is old, and it seems exhausted and in full decline. The essential was laid down in the Primary. Even the most recent groups - those of the mammals and the birds - arose in the Jurassic and flourished in the Cretaceous, i.e., in the course of the Secondary. The entire Tertiary has not seen arise one single type of truly original structure, except perhaps for that of the Gymnophiona amphibians (the burrowing caecilians, which have lost their paws) and the ultra-specialised mammals (the Chiroptera, conquerors of the sky, and the Cetacea, which returned to the sea). Even the Primates, the order to which we belong, appeared before the end of the Secondary. One could practically say: 'We are *all* living fossils'.

THE RELICS OF A SPLENDID PAST

Certain captious and picky researchers have felt that it would be appropriate to reserve the term of 'living fossil' exclusively for *survivors* of *vanished* groups. On this basis, one could legitimately hold that they should, in logic, no longer be of this world...

Let us first of all emphasise that, if survivors of it exist, a group cannot be considered to be extinct. To speak of groups facing extinction is closer to the truth. The 'living fossils' would in some way be comparable to the 'last of the Mohicans'. But, if this is so, would it not also be necessary to consider as 'living fossils' all of the innumerable groups which were substantially more flourishing in former times? For example, to confine ourselves to the Vertebrates alone, *all* of the Agnatha fish, that is, those without jaws, *all* of the Amphibians and *all* of the present Reptiles? These are incontestably three groups which formerly were dominant, but of which (relatively) few representatives remain in our time.

The Agnatha were the masters of the seas in the heart of the Primary, but they were soon replaced in the Devonian by cartilaginous fish, or Chondrychthya (sharks, skates and chimaeras). Today, among the some 20 thousands of species of fish already recorded, there are no more than about 70 species without jaws - an infinitesimal minority! - spread over two large classes: the lampreys and the hagfishes. The present situation is scarcely better with the Amphibians, which dominated the newly emerged lands and the fresh waters of the Carboniferous and the Permian. These potentates of past ages - certain of which, the Stegocephalia, were truly colossal - are represented today only by the salamanders and the newts, the frogs and the toads, which in general are small and shy, and the obscure caecilians.

It is pointless to continue discoursing here on the collapse of the Reptilian Empire, on which the sun never set and which covered, all at the same time, the waters, the lands and the air. All through the Secondary, at the beginning of the Trias up to the end of the Cretaceous, during a hundred millions of years, the Reptiles were the undisputed lords of the planet. And then suddenly - at least on the geological scale of time! - their most prestigious cohorts collapsed, and half of their known orders became extinct: the Plesiosaurs and the Ichthyosaurs in the seas, the bird-hipped Dinosaurs and the lizard-hipped ones on land, and the Pterosaurs in the air. The only ones who escaped from this unbelievable hecatomb were the turtles, the crocodiles, the lizards and the snakes and - which is the most extraordinary thing - an isolated member of the order of Rhynchocephalia, which date from the Primary and which thus precede the spread of the famous dinosaurs. This miraculous survivor is the Sphenodon, the *tuatara* of the Maoris of New Zealand, a large spiny lizard, olive-coloured, and which measures between 50 and 70 centimeters in length.

These Rhynchocephalians were spread across the entire world in the Secondary, at the end of which they appear to have disappeared as if by magic. Today, only a single genus of them is to be found on about thirty small islands scattered off the coast of New Zealand. These islets are located in Cook Strait, which separates the two main islands, South Island and North Island, and in the coastal waters to the north of this latter. There cannot be the slightest doubt that the tuatara is a 'living fossil': it is even one of the most striking examples of such, since its two recognised species are *all* which remains of an entire order of reptiles, which was cosmopolitan in former times but which, with this one exception, is now

Ichthyosaurs by Zdenek Burian.

totally extinct. One can nevertheless raise the question: how does its case differ *fundamentally* from that of some other reptilian groups, which for the most part are extinct?

The order of the Loricata, or crocodiles, and that of the Chelonia, or turtles and tortoises, are almost as old as that of the Rhynchocephalia, since they appeared in the Trias, and thus at the very beginning of the Secondary. The first one, in our days, is now only represented by barely 22 species, distributed over 3 genera, forming 3 different families. The second order still numbers 244 species distributed over 75 genera, forming 13 families. Between the present situations of these three orders of reptiles, the difference is purely quantitative.

If, in all logic, there is granted to all three orders the designation of 'living fossils', then one at once better understands that the same has likewise been done for other orders of vertebrates which are also in clear decline: namely, among the fish, the Polypterus, the sturgeons and paddlefishes, the Lepisosteus or garpikes and the Amiidae or bowfins (all of those which

Darwin had grouped under the name of 'Ganoids'), as well as the Dipneusti or lungfishes (among others, the Lepidosiren, which was also taken by Darwin as an example to define 'living fossils').

Among the birds, one could just as legitimately cite the various forms of more or less wingless runners, which after all are now rare, and which are grouped in the order of the Struthioniformes: the ostrich, the emus, the cassowaries, the rhea and, above all, the mysterious kiwis. Scarcely ten species, distributed over five genera. An infinitesimal residue among the 9,672 species of birds, distributed over 2,057 genera.

And, among the mammals, would it not be in order to place in the category of 'living fossils' not just those with a cloaca, the Monotremes, but also - why not? - all those with pouches, the whole of the order of the Marsupials?

BUT WHERE ARE THE MONSTERS OF YESTERYEAR?

In this respect, plainly dazzled by the general decline in the size of the amphibians and, above all, by the disppearance of the most spectacular reptiles of the Secondary, of which the dinosaurs will forever be the stars, we have somewhat lost sight of the no less impressive crumbling away of the class of the Mammals. Indeed, what remains of the luxuriant variety of the hairy colossi of the Tertiary?

Of the seven recognised orders of Amphibians, four - a little more than half - no longer exist. Among the Reptiles, two-thirds of their orders are presently extinct. But, the losses have not been substantially less among the Mammals, since half of their orders have disappeared.

No more Multituberculata, Triconodonta, Pantotheria or Symmetrodonta: Australia seems to have been deserted by the marsupial lions, the Diprotodontia - sorts of wombats as big as rhinoceroses - and the *Palorchestes*, which had the appearance of an ox with a pig's head and the claws of an ant-eater. Finished also the Taeniodontia and the great Tillodontidae, with their rodentlike incisors, in North America. In South America, what has become of the Megatheriidae, these enormous terrestrial sloths the size of an elephant, and the Glyptodontidae, veritable tanks of flesh and blood. In Madagascar, why are there no longer any lemurs similar to bison or chimpanzee and, in the Far East, where are the gigantic anthropoid apes,

alongside of which the gorillas would cut a pitifully small figure? The great cat is dead. The felines with sabre teeth or dagger teeth have vanished. The knell of doom has sounded for the Protungulates of all sorts: Condylarthra, Notoungulata, Litopterna and Astropothera. Of this considerable super-order there remains only one single Tubulidentate, unique in its genus and even in its order: the *Orycteropus*, the aardvark of the Afrikaners. Where now are the four-tusked elephants, those with curled upper tusks and with lower tusks in the form of ploughshares or shovels or spades? Farewell to the Pantotheria, farewell to the Dinocerata, which, not being content to have three pairs of pseudo-horns, also possessed terrible feline fangs, farewell to the Pyrotheria and the Embrithopoda, like this Arsinoitherium with four horns, of which the first two, united and massive, formed the most formidable weapons for combat. There is no more rhinoceros with a woolly cloak, nor any rhinoceros without horns, nor chalicotheria, which resembled a horse with the claws of a predator and the silhouette of a hyena. No more baluchitheria either, the greatest of all land mammals, which exceeded five meters at the withers and which could raise its head to a height of more than eight meters - men could have walked under its belly without having to bend over! And, one may shed a tear for the wild boars as big as hippopotami, and whose tusks measured around 60 centimeters, and for the proto-stags, which had a second pair of horns on the nose, and for the superb Irish elk, whose antlers could exceed four meters in breadth. All of these war machines have been cast aside.

In truth, the last giant mammals surviving today - whales, elephants, rhinoceroses, hippopotami, large felids and anthropoid apes - all are at the edge of the abyss. That man is proliferating at this moment in an explosive - and thus suicidal - fashion should not conceal from us an important piece of evidence: like the majority of the other classes of vertebrates (the two groups of fish without jaws, the Amphibians and the Reptiles), the Mammals are in decline. It is the twilight of the nipple-bearers. Why then would one refuse them the enviable title of 'living fossils'? Let us not forget that it

has also been conferred on numerous classes of invertebrates, for the most part extinct: Monoplacophoran mollusks (*Neopilina*) and the Amphineura mollusks (chitons), the Onychophora (velvet worms), the Xiphosura (king crabs) and the Crinoidea (sea-lilies). It has even shamelessly been given to certain entire phyla: that of the Brachiopoda and that of the Pogonophora! Both of these, however, still contain numerous living representatives: the Brachiopoda, with some 250 species distributed over 12 genera, forming 7 families and 2 orders, and the Pogonophora, with 43 species distributed over 11 genera, forming 5 families and 2 orders as well. And, there seems to be little doubt that these two phyla are still cosmopolitan. The Brachiopoda or lampshells are found in all of the seas, from the low tidal mark to the edge of the continental shelf. As for the Pogonophora, they carpet the bottoms of the greatest depths, and in sometimes unbelievable quantities.

THE QUARREL BETWEEN THE ANCIENT AND THE EVEN MORE ANCIENT

If, for a zoological (or botanical) group of any size whatever, admission to the club of 'living fossils' would depend in essence on the more or less elevated percentage of surviving species in relation to the total number of species having existed in the past (in any event unknown and, moreover, unknowable), then it must be concluded that there are variable degrees in the statutes of the association in question: one could be *more or less* a 'living fossil'. It is as if one said that a fossil being could be more or less living, or that a living being could be more or less fossil. Such a conclusion is worthy of Pierre Dac or Groucho Marx.

One ends up in another blind alley if one endeavours to judge the 'living fossils' according to their degree of antiquity. As is known, they exist in all ages.

The most archaic are the Radiolaria, these single-celled organisms covered with a siliceous shell, sculptured like a jewel of Toledo, the earliest remains of which have been found in the Precambrian, and thus before the Primary Era, in layers dating back to well over 500 millions of years ago. Even in our days they are still so abundant that their skeletons, accumulating at the bottom of the sea, create there a special sand which is called radiolarian

mud, and which covers immense areas in the tropics. This is also the case with other Protozoa, the foraminifera, which are protected by a calcareous shell, and, in particular, the largest among these, the nummulites, certain ones of which can reach a diameter of 4 centimeters. These nummulites have always been so numerous, since the Precambrian, that their remains ended up by forming the calcareous stones with which were built, among other things, the pyramids of Egypt. From this same Precambrian there date also the first bacteria, algae and sponges. Really nothing other than very ordinary.

From the Primary or Paleozoic come to us certain species of bacilli, still widespread in our days, such as *Bacillus circulans* or *B. amylobacter*, and several genera of living beings which are still quite active today: the indestructible tree of China, the *Ginkgo*, and, it is believed, the octopus-like plant *Welwitschia* of the Namibian desert, five forms of Brachiopoda, including the famous *Lingula* and a whole series of Mollusca. Among these there may be cited the monoplacophoran *Neopilina*, practically identical to the *Pilina* of 500 millions of years ago, the, gasteropod *Pleurotomaria*, several lamellibranchia, like *Arca*, *Nucula* and *Avicula*, and, finally, the cephalopod with a spiral shell and capable of swimming, the *Nautilus*. This is already more impressive.

A multitude of recent genera go back to the Secondary or Mesozoic, among others a new trio of Brachiopoda, the ancient *Limulus* and the peracaridan crustaceans, such as *Phreaticus*, among the Arthropoda, the sea-lily *Pentacrinus* and five forms of urchins among the Echinoderma, as well as a whole range of very old fish. First of all, there are those without jaws, the lampreys and the hagfishes, and then, among the cartilaginous fish, some sharks, such as the frilled shark *Chlamydoselachus*, which resembles like a brother the *Cladoselache* of the Devonian, the pig-shark *Heterodontus*, well known from Port Jackson, in Australia, and even the vulgar six-gilled shark (*Hexanchus*). Among the bony fish we must not fail to mention those with lungs, called Dipneusti, that is, 'with double respiration', and, above all, the, famous fish with paws, *Latimeria*, of the family of the coelacanths, already present in the Cretacean in a very similar form, *Macropoma*. And, let us not forget, among the amphibians, the frog *Leiopelma*, nor especially, among the reptiles, the *Sphenodon*, the last of the Rhynchocephalia, and almost identical to the *Homoeosaurus* of the Jurassic. A truly spectacular-collection.

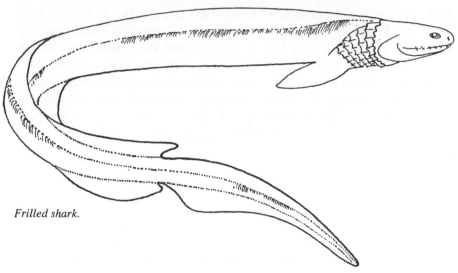

Frilled shark.

This being said, are the genera of the mammalian 'living fossils' of the Tertiary or Cenozoic any less striking? Like the enigmatic South American marsupials of the genus *Caenolestes*, astride the two major sub-orders, the extremely rare giant insectivore of the Antilles, the almiqui or *Solenodon*, the various armadillos with their strange armour, and the minuscule jumping primate, with eyes literally larger than its belly, *Tarsius*, the ghostly tarsier, all go back to the Paleocene, or some 65 millions of years ago. Only slightly less venerable are the sarigues or opossums, whose fur adorns many elegant but unfeeling ladies, the marvelous cohort of the lemurians and even the vulgar dormice: they all come from the Eocene, namely, from about fifty millions of years ago. The little shrews, the bats, these oddly befurred flying creatures, many monkeys of the Old World and rodents like the squirrels and the jerboas, and - which is somewhat less astonishing - the pangolins, these odd mammals covered with scales, all are relics of the Oligocene, which began 35 millions of years ago. Animals as familiar as the moles and the hedgehogs of our gardens, the cats of our homes, the bears, represented in our cradles, the hares of our fields, the rats of our sewers, all go back to the Lower Miocene, whereas the generally more exotic animals, such as the phalangers, the ant-eaters, the ancestral dogs, the hyenas, the rhinoceroses, the wild boars, the hippopotami, the giraffes and the okapi appeared in the Upper Miocene, between 25 and 35 millions of years ago. Among the most recent arrivals, in the Pliocene, between 5 and 25 millions of years ago, one may cite the macaques and the langurs, as well as the elephants, the tapirs and the horses. Concerning the genus *Homo* itself, with its 2 to 3

millions of years of age at the least, one can truly not say that it was born yesterday.

In summary, all of the geological eras have left behind more or less numerous relics of their characteristic fauna. Among them, why would the most ancient ones not merit more than the others the title of 'living fossils'? And beginning from what degree of antiquity? Where the devil can one trace a line of demarcation which makes some sense in the vast range of forms dating from thousands, hundreds or tens of millions of years ago, or even from only a few millions of years ago?

Nine-banded armadillo (Photo Karl Marslowski - Rapho).

IN SEARCH OF AN IDEAL

In order to conserve for the 'living fossils' the exemplary character which ordinarily cannot be imparted by a more or less exceptional nature, we have striven to build into their definition various clauses which are aimed at reducing considerably their number. These limitations are aimed at both the factor of *space* and at that of *time*.

What qualities were required in general of a 'living fossil'? A thoroughgoing analysis of the literature shows that, aside from relative age, several quantitative measures have been taken into consideration in order to justify applying the term of 'living fossil':

1.   the grade - as low as possible - of the form in question in the natural hierarchy, and, ideally, that of the species;

2.   the number - as small as possible - of species which represent it at the present time;

3.   the number - to the contrary now as high as possible - of species which formerly composed the group to which it belongs;

4.   the present range of distribution - as limited as possible - of the form in question;

5.   the old area of distribution - to the contrary now as broad as possible - of the group from which it came;

6.   the grade - as high as possible - of this group in the natural hierarchy.

In short, the *nec plus ultra* of the 'living fossil' would be a unique species in its genus, surviving only in an extremely limited region, and constituting all of that which remains in our days of a vast and formerly flourishing group: a phylum, preferably, formerly spread across the entire world and this, if possible, in the course of the Primary Era, since the largest part of the diversification of the Animal Kingdom as well as of the Vegetable Kingdom has taken place since that epoch.

68

This is almost asking the impossible. The fact is that none of the forms which have been termed 'living fossils' fulfils all of these conditions. For most of the cases, it is a matter of relatively recent forms (like the aardvark, *Orycteropus afer*, sole present representative of the superorder of the Protungulates, but which goes back only to the Miocene), or of forms which are very ancient but which are still widely diversified and fairly common (like the Brachiopoda and the Crinoidea, both dating from the beginning of the Primary, but which are still cosmopolitan in our days, or, again, like the king crabs and the neighbouring Cephalocaridea, which also have come down from the Primary, but each represented by at least two or three genera, scattered over most of the oceans). It was sometimes also a matter of assuredly very ancient forms, but forms not belonging in any way to a decadent group (like the *Pleurotomaria*, a triassic genus of the still quite prosperous class of the Gasteropoda, which gathers in all of the snails, slugs, winkles and whelks of the planet), or else incontestably unique forms, but which belong to small groups already hidden in time and, moreover, from an obscure past (like the species *Welwitschia mirabilis*, or the two other genera of Gnetale plants). There were even forms like the hoatzin bird, which only had an 'archaic appearance'.

THE CREAM OF THE 'LIVING FOSSILS'

In the way of 'living fossils' which most closely approach the ideal type, one is reduced to citing only the following:

1. the genus *Sphenodon*, the two species of which, distributed over the different small islands of New Zealand, are strictly the present residue of the reptilian order of the Rhynochocephalia, which prospered all through the Secondary;

2. the species *Latimeria chalumnae*, the only one which survives in our days from the Devonian order of the Crossopterygian fish, and which seems to be domiciled around the Comoros and in the Mozambique Channel;

3. the species *Platasterias latiradiata*, last relic of the sub-class of the Somasteroidea, aberrant starfish which had been believed extinct since the Ordovician, in the Primary, and which appear today to frequent only the waters of the Pacific Coast of Central America;

4. the genus *Neopilina*, the few rare species of which, discovered in the depths of the Pacific Ocean off the coast of South America, from Panama to Chile, are the sole survivors of the archaic class of the Monoplacophoran mollusks, which date from the Carboniferous, toward the end of the Primary;

5. the genus *Nautilus*, represented at the present time by four or five species scattered across the Pacific and the Indian oceans, from Madagascar to Polynesia, and constituting the tiny residue of a group of cephalopods with nacreous shells, which counted more than 2,500 species in the course of the Primary, from the beginning of the Ordovician to the end of the Silurian;

6. the species *Ginkgo biloba*, the sole and unique species remaining of the plant group of the Ginkgoales, born in the Primary and having known its apogee in the Jurassic, right in the middle of the Secondary.

Quite evidently, certain purists could demand an even more severe selection and, for example, could claim disqualification of those forms which count several species. This would reduce the total number of the least contestable 'living fossils' to three, a figure which, moreover, could well decrease even further if, as is quite possible, there should be discovered another species of coelacanth. But, even if one holds only a half-dozen forms of 'living fossils' to be beyond discussion, one will still be exposed to the criticism of not having included among them certain of the most classical examples of the matter, like the Brachiopoda, the *Peripatus* and the king crab, but above all for having eliminated even those which Darwin had cited precisely for the purpose of creating the expression: the Ganoids, the *Lepidosiren* and the platypus. This, it must be recognised, is unacceptable.

All of the efforts made to limit the number of 'living fossils' by means of an interplay of restrictive criteria have been doomed to failure. Thus, in 1967, plainly moved by the sharp criticism made of the expression, which to say the least is an unfortunate one, by myself in *Sur la Piste des Bêtes Ignorées* (1955), a student at the University of Paris, Serge Cassagne, published a laboratory study in paleontology entitled *Analyse de la Notion de 'Fossil Vivant' (Analysis of the Concept of 'Living Fossil')*. Here is the convoluted definition which he was led to propose after a systematic and truly painstaking study: 'One will term 'living fossils' those present species which can be classed among the fossil taxons of rank inferior or equal to the super-family and of age anterior to the end of the Miocene'.

Why 'the super-family' and not 'the family' or 'the order'? Why 'of the Miocene' and not 'of the Oligocene' or 'of the Pliocene'? One cannot imagine a more sparkling confession of an absolute arbitrary.

EVOLUTION TAKES PLACE THROUGH BUSH-LIKE BRANCHING OUT

We must make up our minds: *every* animal and *every* plant of our time is a 'living fossil', or else 'living fossils' do not exist. This goes along with the fact that this idea itself is today obsolete: it no longer makes any sense in the present state of our knowledge concerning the evolution of the whole of the living world.

Well before the idea of such an evolution appeared, there was more or less a generally held belief in a sort of *scala naturae*, in other words, a scale or, if one prefers, a stairway of Nature, leading from the most simple and the least organised living forms up to the most complex and most refined forms, of which Man, to be sure, was the crowning glory. This concept of a 'great chain of beings', as it was so well named by Arthur O. Lovejoy in 1942, was founded originally on a very anthropocentrist religious belief, according to which Man would be not only the latest product of a creation, but its very purpose. When, with Lamarck and Darwin, the transformist theories came little by little to be accepted, Man, in the same perspective, was automatically considered as the summit no longer of divine creation but of biological evolution. This skilful transposition had the advantage of being accepted without too much aversion, as it caused the least possible disturbance to established ideas. Nevertheless, there was nothing scientific about it, since the premises of the syllogism which culminated there were purely dogmatic and founded on faith.

If, in our days, the evolution of the living world is represented graphically, the result does not at all resemble a scale, a flight of stairs or a chain. Rather, it resembles a genealogical tree or, better yet, a thick bush. The new forms, in fact, do not evolve from ancient forms which thus disappear through transformation: they diverge little by little from existing forms, which survive while still pursuing their own development.

A truly original type of structure is never the culmination of a high degree of specialisation. It does not arise at the extremity of one of the branches of the bush which represents the expansion of an earlier group. Rather, it always stems from the type which is synthetic, undifferentiated and malleable, 'good for all purposes', which is found at the base of the said bush. After having pursued an uneventful evolution in the quiet shadow which is assured by its generally small size, its commonplace aspect and its unspecialised habits, the new type, seeking to find its way, ends up by finding the conditions which are the most favourable for its expansion. At this moment it suddenly gives rise to an almost explosive efflorescence of the most varied forms. From the lower part of this still green bush, while increasingly more extravagant forms are becoming differentiated, another type, humble and diffident, can break out, which one day will assure the apotheosis of another completely original group. And so on.

Dolichosoma (left) by Zdenek Burian.

In brief, the evolution of a new type of structure always occurs in parallel with that of older types, but with a certain time delay. In the genealogical tree of Life, the various groups are joined only at their bases. The animal empires do not succeed one another by replacing one another: they step on one another and come into combat where they are in competition to occupy the same habitat. Yesterday, it was thought that forms were replaced through metamorphosis. Today, it is known that they are superimposed on one another and that, on occasion, they destroy themselves mutually to the point of extermination, or nearly so.

Those which are called the 'living fossils' are the beings which have escaped from these confrontations, persecutions, banishments and massacres, either because they never encountered adversaries of their stature, or else because their territory has not yet been disputed, or else because they have good passive or active defensive arms, or even because, quite simply, they are well hidden. All methods are valid for survival, and all are used.

SURVIVAL IS THE RULE, DEATH THE EXCEPTION

So this is why there remain, in our days, representatives of all of the zoological phyla (we can forget here the vegetable kingdom) which have developed on our planet since the birth of life, even those which are held to be the most 'primitive', the most 'archaic' or the most 'inferior', if that means anything from the biological point of view.

There was earlier believed to be one single exception to this rule: the enigmatic Conodonts, the fossilised remains of which, in the form of rods bristling with denticles, have been found in numerous rocks dating from the Primary and the beginning of the Secondary. In the course of the last two decades they had been classified among the plants, the Coelenterata, the Aschelminth worms, the Gnathostomulids, the Aplacophoran mollusks, the Tentaculata, the Chaetognatha and the Chordata. It would almost be simpler to list the phyla in which efforts have *not* been made to place them. Finally, in 1988 there appeared an impressive monograph by Walter C. Sweet, according to which the Conodonts would constitute an entirely distinct phylum, extinct for 200 millions of years. However, in 1992, more extensive research revealed that they are, in reality, the teeth of an organism related to the hagfishes, little agnathous fish, and that consequently they must be placed at the base of the Vertebrata.

73

In truth, survival, even very partial, is everywhere the rule rather than the exception. At the level below that of the phylum, there are only very few broad zoological groups which have disappeared entirely. We may judge from the following: of about seventy classes of generally recognised animals, there are only three for which descendants have not been found in the present epoch:

1. the Archeocyatha of the Cambrian, of the most doubtful of affinities but related vaguely, it seems, to the foraminifera, to the sponges and to the cnidarian coelenterata, i.e., to the most formless animals.

2. the Bellerophontidae, these little mollusks from the second half of the Primary, the shell of which is coiled in a symmetrical spiral in a single plane, like that of the nautilus, but in only faintly outlined fashion.

3. and, last but not least, the prestigious Trilobita, these crustaceans which literally swarmed over the bottoms of all of the seas of the world from one end of the Primary to the other.

Why these few rare classes - as a matter of fact, exceptional classes - apparently became extinct no one knows for sure. It is even more difficult to explain why many zoological orders appear to have disappeared, whereas others which are related to them are still flourishing. To illustrate this, it is enough to review the plethora of 'scientific' theories - which indeed are often contradictory - which have been proposed to account for the supposed extinction - which in any case is generally accepted - of the dinosaurs, the latest one in fashion, and which is nonetheless well established, being a shower of comets. One would say that none of the men of science who have set forth such lucubrations had, first of all, posed a simple question: 'How the devil could any sort of catastrophe selectively destroy a particular group of animals and leave unharmed all of the others sharing the same habitat, even their closest relatives?'

The strict truth is that, like individuals, groups of animals of any size - from the simple species to the phylum - quite simply die 'of old age', namely, as a result of the progressive accumulation of ailments which are generally benign for the individuals, and from various factors which are unfavourable for the species or the higher categories. A minor accident, like a fall or a congestion in the case of the former, and a more or less extensive natural

catastrophe in the case of the latter, often suffice - like a veritable *coup de grâce* - to put an end to the functioning of an organism or to the potential adaptation of a form of structure which is worn out, fatigued, and at the end of its run.

THE GRAND ILLUSION OF 'LIVING FOSSILS'

The difference is that, for the individuals of each species, the potential longevity manifestly is programmed in their genes, whereas for more extensive groups the duration of existence - at least until we know more - appears to be more variable. When they are particularly well adapted to their environment, certain zoological communities succeed in surviving longer than others, as Darwin was careful to point out. New adaptations becoming less and less necessary, their evolution slowed down to the point of appearing stationary, and it could be said that their existence might be extended indefinitely if no major upheaval were to take place in their surroundings. This would lead one to believe that, in such cases, the progress of evolution stopped early, and so they have come through to us from the very depths of the mists of time.

Nevertheless, it is rather difficult to assert with confidence that a particular form of structure has not changed for millions of years. For, ultimately, what do we know of the beings of former times? As has already been emphasised, we never possess anything more of them than scanty debris: a shell, a skeleton, or even a poor fossil outline. On the basis of these meagre remains, can one be sure that the being to which they belonged has undergone no substantial and detailed modification, from the cellular, anatomical or physiological point of view, before coming to us with apparently similar appearance? The installation of a thermal regulation mechanism, the development of venom glands or lungs, the construction of a heart with perfect circulation, the avatars of the kidney, the delicate refinement of an endocrine equilibrium, the regulation of an immune defence system, the growing complexity of the sensory or motor organs, the more highly developed coordination of nervous connections: so much important progress, among much other, which leaves scarcely no trace on the hard parts of the organism. So this is what makes so ordinary that which has been called, no doubt abusively, 'living fossils'.

In any case, the omnipresence and the commonplace nature of these latter should enable us to sweep away one of the objections most frequently raised to certain claims of cryptozoology. The possible survival into our times of supposedly extinct forms of life - whether these be relatively recent prehistoric men, archaic cetaceans, dinosaurs, giant amphibians, fish with paws or trilobites going back to the most ancient times - should surprise only those backward souls who still bow down before an outdated concept of biological evolution.

Traditional sea-serpent after the Journal des Voyages (1914).

5

THERE ARE LOST
WORLDS EVERYWHERE

'There doubtless exist, in regions of the world which are still inaccessible, as well as in the great depths of the sea, creatures of which science
will be unaware for a long time to come.'

(The animal collector Joseph Delmont, *Vingt ans autour du monde*
[Twenty Years Around the World], 1932)

I t is now known that discoveries are continuing ceaselessly to be made
of animals the size and appearance of which exclude absolutely their
being able to pass unnoticed. Moreover, the conviction has arisen that
these animals can perfectly well belong to forms dating back to the ages -
even the most ancient - of the development and diversification of life. The
final question which is then to be posed is, of course: where the devil could
such species still be hiding?

The response which comes immediately to mind is also quite evident. It
is in the last *terrae incognitae* of the globe. But, do there still exist today
blank spots on the map of the world?

In an epoque when Man has undertaken to explore the other planets of the
solar system and is preparing to venture into interstellar space, it is generally
considered that our Earth is known thoroughly, from top to bottom.

Above : Chlamydosaurus.

This idea is deeply rooted in popular thinking. Intrepid men have travelled through all regions in all directions: they have even reached the two ends of our world - its frozen poles. The airplane enables us without difficulty to fly over regions reputed to be impenetrable. Aerial photography has resolved all geographic and cartographic problems. Everest has been vanquished, and the Mindanao Deep has been probed. So now, where could one still turn up the substance of marvels, slake one's thirst for mystery, discover new enigmas to be unravelled, if not in new worlds - inviolate, extra-terrestrial? One might even say that Man has rejected his planet like an old toy in which he is no longer interested.

But, in reality, we are far from having exhausted all of the resources offered by our terraqueous globe. It is still true that there are more things in heaven *and on earth* than were dreamed of in the philosophy of Horatio.

To have no interest in our planet? Our knowledge of it, when all is said and done, comes down to the results of an enormous amount of surveying work. We know the form and size of the Earth, the limits and the topography of its continents, the characteristics of the vegetation which covers them here and there; at certain selected spots we have even scratched the terrestrial crust to study its composition, as well as the fossils which it contains. We know generally the topography of the ocean bottoms and the great currents which flow across them. But, imagine the case of a biologist who succeeded in developing a precise and accurate portrait of a human being: could he, without giving way to presumption, claim to have perfect knowledge of the being's anatomy, its physiological troubles, the condition of its tissues and its pathology?

Above all, we must deal with the matter of the enormous work involved in reducing and wiping out the blank spaces of the terrestrial planisphere. Just in the middle of the 20th century, an areopagus of French explorers endeavoured to set up a balance sheet of what remained to be done to complete the conquest of all lands above sea level. At that moment the extent of our ignorance was still quite impressive. Apart from the immense Antarctic region, as large as Europe and the United States combined and two-thirds of which were still unknown, eight notable 'blanks' were still scattered across all of the continents with the exception of North America, and were even found on certain large islands, notably Greenland and New Zealand.

These blank spots were destined to shrink like a piece of shagreen, and even to disappear, before the onslaught of expeditions of exploration, and even more so through the progress of aerial cartography. Over the course of the years one was soon to pass from sequences of regional photographs taken at high altitude (30,000 meters) to global photographs made by satellites, and then to photogrammetry, which is based on stereoscopic effects and thus yields a relief map.

In a fine work entitled *Cartographie: 4000 ans d'Aventures et de Passion (Cartography: 4000 years of Passion and Adventure)*, Thierry Lassalle once again, in 1990, summarises the question quite well. Rejoicing with the poets 'who believe in the need to dream', the author indeed felt called upon to note that 'as by a miracle, a part of the mystery remains: from the heart of the island of Borneo to the south of the Algerian desert, and passing through the frozen domains of the poles, thousands of square kilometres continue to hold their secrets.'

We should note in passing that these empty spaces which, as Lassalle, spokesman of the Institut Géographique National (National Geographic Institute), says, 'still escape the precision of the cartographers', do not all correspond to the last unknown areas identified by the French explorers at the middle of the 20th century. In other words, the 'blanks' of some are not necessarily the 'blanks' of others, which is perplexing, to say the least.

The evidence must be faced: in spite of all the technical progress so far, at the present time there still exist spots of absolute whiteness on the map of the world. This, of course, stems sometimes from the fact that there are countries whose air space cannot be violated with impunity. And besides, even when a region may have been mapped perfectly, it could still conserve a part of its mystery. The geography of the zoologist, of the ethnologist, of the botanist, of the paleontologist or of the archeologist is not necessarily that of the cartographer. The most sharp-eyed aerial cameras, whose resolving power can reach 50 cm, still do not succeed in piercing the screen formed by the canopy of the great forests, nor even the mirror of murky waters.

A TOUR OF THE WORLD OF THE LAST HIDING-PLACES

Cryptozoology is devoted to hidden animals. Thus, it is not with the eye of the cartographer but, rather, with the eye of the zoologist that we must again review and examine the various regions of the planet. The work of exploration by the zoologist often only begins when the topographic drawings of the cartographer are completed. What interests the cryptozoologist is not the distribution of the sometimes immense zones which are designated laconically on maps by terms such as 'hygrophilous equatorial forest', 'desert with xerophilous vegetation', 'marshes', 'mangrove', 'steppe', 'savanna', 'tundra', 'taiga' or 'thorny underbrush'; rather, it is what is hiding *in* these zones.

We can pass fairly rapidly over the regions around the poles, both north and south, however little known they may still be. The Antarctic, Greenland and the innumerable islands which dot the frozen seas are scarcely ever trod by human feet, but all of these regions often are flown over by airplanes, sometimes at low altitude. There can be seen quite clearly the seals taking their sunbaths, the herds of musk oxen, the populations of auks or penguins, as the case may be, or the polar bears, which sometimes are encountered hundreds of kilometers inland. On this gleaming background of snow and ice, nothing escapes being seen, any more than a flea on a white sheet. So, here we have the justification for the point of view expressed by Paul-Emile Victor, the great explorer of polar lands, when he writes: 'Perhaps some day in these regions there will be discovered still unknown species of animals. But, it must be understood: they will be only animals of very small size, for example, insects.'

Even though vastly better explored in all senses, Africa offers many more resources to naturalists seeking to discover animals of a more appreciable size. Among the regions which may still harbour such creatures, there must be cited first of all the great equatorial rain forests which girdle the dark continent almost completely, from Liberia to the zone of the great lakes, a spread of over 4000 kilometers. It is to this band of impenetrable vegetation that we owe five of the greatest zoological discoveries of this century. How could this be at all surprising? The trees there grow to heights up to 60 meters, and their branches intermingle so closely that they form a canopy which can scarcely be penetrated by the sun's rays. This roof of vegetation

Facing page: Plesiosaurs by Zdenek Burian.

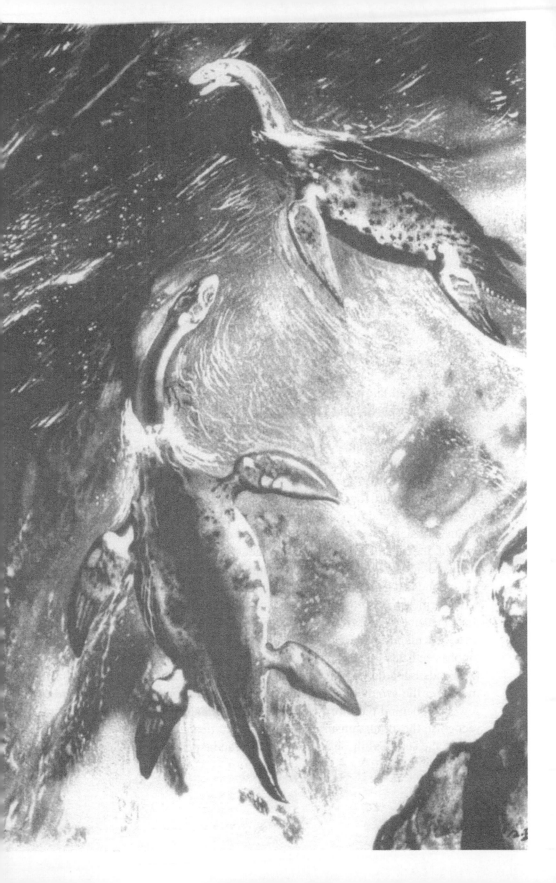

literally holds in the heat and makes of the forest an oven, both in the day and in the dark of night. In some places it is even partly flooded, so that one can move only by wading in water up to one's waist. Pirogues cannot be used because of obstacles of all sorts: trees, underbrush and lianas. Then, going farther to the east and to the south, there should also be mentioned the mountain masses of Ethiopia, the mountain forests of Kenya and Tanzania, as well as all of the marshy areas and regions studded with lakes, which can shelter unknown beasts of considerable size: in the Sudan especially, where the immense marshes of the Adar and those, even larger, of the Bahr-el-Ghazal extend over thousands of square kilometers, as well as those in Katanga and in Zambia, which are scarcely less impressive.

Illumination does not necessarily come to us from the Orient. In Asia, the gigantic tentacular network of mountain chains which surround the heart of the continent has lost but little of its mystery, in spite of spectacular exploits on the part of a number of mountaineers. As for the Siberian taiga, on which the sun never sets, its unsurpassed size is the surest guardian of its secrets. But it is the tropical jungle, including sometimes that of countries held to be well known, which, in these cases as well as in Africa, remains the inexhaustible treasure for cryptozoologists. The jungles of Assam, Myanmar (ex-Burma), southern China, Indochina and the Malaysian peninsula and, beyond Singapore, those of Borneo and Sumatra, remain rich in possibilities.

Still farther toward the east, the situation is the same in the tropical forests of the Philippines and of Sulawesi. In New Guinea, to be sure, one can move about more easily today than fifty years ago. In Irian Jaya, the western half of the island, air services connect the various Christian missions scattered across the entire country, and which are under the protection of the Indonesian government. There can be reached in this way places which were formerly held to be totally inaccessible. But, this in no way means that one can explore without difficulty what is the largest island in the world after Greenland. The Belgian naturalists André Capart and Xavier Misonne emphasised this quite well in 1974, following the scientific expedition of the Fonds Leopold III in Irian Jaya: 'one must have tried to go a few dozen metres in the swampy forest, and one must have tried to move forward a few steps behind the thorny curtain of the highaltitude forests, in order to recognise the almost unsurmountable difficulties which must be overcome before the secret of these ageless worlds can be penetrated'.

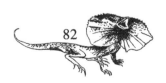

One truth must be recognised: in regions which have been little explored, the arrival of air transport, far from improving traffic on the ground, has had the very opposite effect of inhibiting it, in that improvements in the road network then are no longer seen as a priority, or even as merely a useful thing.

In this regard, what can be said of Australia, this continent whose interior is for the most part covered with deserts of sand and rock or thorny brush? The regions which are the least hostile to penetration are those of immense steppes covered with tall grass and over which are scattered stunted bushes. Only a few hardy prospectors, driven by the hope of discovering some elusive vein of gold, occasionally venture into these unknown zones which most people prefer to cross by airplane.

The continent of mystery *par excellence* remains without a doubt South America. Bernard Flornoy, who passed a good part of his life exploring in Amazonia, speaks of it as an 'immense night of trees which extends over five millions of square kilometres'. How can we explain that four centuries of investigation have not been able to strip this basin of all of its mysteries? According to the French explorer, 'it is because the virgin forest constantly sets its irresistible power against the efforts of men, a power which, in the space of a few hours, transforms all that which has rotted into a triumphant new life. A path is opened, and it is wiped out immediately. And, if you fly over the Amazonian forest in an airplane, cities like Belem and Manaus in Brazil, and Iquitos in Peru, have the aspect of clearings. As for the clearings and glades where the Indians themselves live, they are invisible.'

Since Flornoy wrote these lines some forty years ago, many calamities have befallen these regions. In the very heart of Brazil, Brasilia, the federal capital, has been built, a mad dream which has become a nightmare. And, more or less everywhere, there has begun the systematic destruction of the great tropical forest, partly for the needs of the lumbering industry, but primarily to obtain wood simply for fuel and to create in the cleared zones vast areas given over to agriculture and stock-raising. Bulldozers and chain saws early on were managing to clear away some 40,000 square kilometers of forest each year. At this rate, being nibbled away on all sides, the Amazonian forest is threatened with disappearance within the coming decades, which would amount to an ecological catastrophe without precedent, not only for the local fauna and flora but for all of humanity as well.

83

In the meantime, however, in the interior of the shredded pieces of the dense forest, in the depths of which the Indians, the only true aboriginals, are being mercilessly massacred or condemned to famine, immense marshes surrounded by low dikes of earth, and covered overall by a tangle of luxurious vegetation, remain the ideal refuge of animals and plants of all sorts. A dense mass of concentrated life: the biosphere in all senses of the term. A teeming world of living species which will die out along with their substratum when this also disappears, doubtless even before having been discovered by western science.

THE 'TEPUIS', ISLANDS IN TIME

In South America, one corner appears to merit especial attention on the part of all who are interested in cryptozoology. Not only because it is suspected of harbouring enigmatic 'little men' and the most bizarre sorts of monsters, but because what it contains represents in a certain way the very archetype of the inaccessible natural reserve. This privileged corner is located in the southern part of Venezuela, where the numerous sources of the Orinoco are located, and which long remained undiscovered.

There, in the *Gran Sabana* to the east - an almost impenetrable jungle - but also in the *Territorio Federal Amazonas*, there are to be found rising up here and there gigantic *mesas* (plateaus) of sandstone, with vertical cliffs and level summits, the height of which varies between 1,000 and 3,000 meters. In the east, this type of truncated needle is called a tepui (that is, 'dwelling of god', in a dialect of Caribbean origin, as one says *tipu* in the neighbouring Guianas); in Venezuelan Amazonia, one speaks more often of *cerro* (i.e., 'mountain' in Spanish), but also more specifically of *jidi*.

In his description of the Empire of Guiana, Sir Walter Raleigh (1552-1681) gave an admirable description of these tepuis, on the basis of information obtained from the Indians, portraying them as enormous fortified castles with vertical walls and flat roofs. But, it was not until 1838 that the first of them, Roraima, 2,810 meters in height, was discovered by two German explorers, the Schomburgk brothers - Robert Herman (1804-1865) and Moritz Richard (1811-1891). As it appeared to them utterly impossible to climb, they limited themselves to travelling around it for about a month, in order to situate and describe it accurately. To be sure, their disclosures incited many other travellers - zoologists, botanists, geologists and simple

alpinists - to try to climb it, but always in vain. It was only in 1884 that the anthropologist Everhard Ferdinand Im Thurn (born in 1852) and his colleague Perkins succeeded in climbing up Roraima to a height of 2,365 meters. At that point, alas, at less than 500 meters from the summit, a large crevasse barred their way and prevented all further progress.

The thoughts which Roraima had inspired in Im Thurn before he attempted his ascension are of capital importance, for they summarise a situation which was to become mythical: 'Assuming that the summit is truly inaccessible, not only to men but also to all animals not having wings, there are some who claim that on this plateau, cut off as it must be from all communication with the rest of the world, it is quite possible that animal forms of a primitive type are to be found, forms which have not undergone the slightest change under the influence of newcomers from the outside, since the moment when the fragment of the plain first found itself in suspended isolation between earth and sky. Whether the place is completely sheltered from such modifying outside influences or not, it is at all events certain that not only its fauna but also its flora must exhibit characteristics of great interest.'

What Roraima really contained was to be learned only in fairly recent times, when scientific missions began methodically to explore its summit, beginning in 1926 in particular. In the meantime, other tepuis had been discovered in the same region: ultimately, hundreds of them were found, but it turned out that there were only about a hundred of them whose heights exceeded a thousand meters. Most of them are quite impossible to climb, but in recent years the use of highly manoeuverable helicopters has enabled the curious to land on their summits without much difficulty. All of the tepuis appear evidently to be the scattered fragments of a vast sedimentary shield, which little by little was cracked, broken up and disintegrated by erosion. When did all of this happen? This is known only very approximately. Geologically speaking, this could have been in the Tertiary, and thus relatively recently, but it is not impossible that the tepuis date from the Cretaceous, i.e., from the end of the Secondary or, in other words, from the time of the dinosaurs and other fantastic reptiles.

However this may be, when the exceptionally strict isolation of what the German naturalist and explorer Uwe George (born in 1940) was to call poetically *Inseln in der Zeit* (Islands in Time) became known, it quickly fired the imagination of many. In any case, this idea had led in 1912 to the

publication of a famous adventure novel, *The Lost World*, by Arthur Conan Doyle.

As the reader will no doubt recall, in this novel a group of British explorers in South America, led by the effervescent Professor Challenger, discovered and finally climbed a high plateau, which was completely isolated from the surrounding regions. There, they discovered, among various relics of the past, a good part of the prestigious reptilian fauna which dominated our planet in the Secondary, as well as sabre-tooth tigers and even some ape-men. As L. Sprague de Camp and Willy Ley emphasised in 1952 in their book *Lands Beyond:* 'The life forms, evidently assembled for dramatic effect only, originally existed in at least three continents and four geological periods'. In the face of the quite understandable incredulity with which the report of these travelers was received upon their return to their home country, the expedition chief opened a wooden packing-case, out of which flew a live pterodactyl - an argument difficult to refute, it must be admitted. In the first cinema film made from this novel, the story was considerably enhanced: instead of the pterodactyl, it was a giant diplodocus, brought back alive to London, which then broke its bonds and got loose, sowing terror throughout the City and then, finally - and fortunately - diving into the Thames and escaping.

It is generally believed that it was Roraima which served as a model for this story by the father of Sherlock Holmes. In reality, this was not the case at all, except perhaps indirectly.

THE ILLUSION OF 'LOST WORLDS'

Exploration of the flat summits of the tepuis by naturalists of all stripes would soon show that the dreams of Conan Doyle were inordinate and exaggerated. Neither dinosaurs nor plesiosaurs nor pterodactyls nor pithecanthropuses were found there. On the other hand, apart from an extraordinary vegetation, in large measure unique and of astounding diversity, as well as thousands of unknown species of insects and arachnids, which are necessarily found in any part of a newly explored tropical forest, there were found there a whole range of relict birds, some mice, and marsupial shrews of the genus *Caenolestes*, as well as some striking new forms of reptiles and amphibians of small size: little lizards and minuscule toads and frogs. At the very most, an octogenarian Latvian explorer, Alexandre Laime, claimed to have seen

Facing page: Plesiosaurs by Zdenek Burian.

on Auyan Tepui, in 1955, large aquatic saurians, but whose length in any case did not exceed 1.20 m.

But isn't that just about what any competent cryptozoologist should expect? The size of large terrestrial animals is always proportional to that of the substratum on which they live. This is, in fact, a strictly mathematical question: their territory must be able to feed them. This, for example, is why islands often harbour dwarf versions of continental forms. Now, the most monumental tepuis - that of Auyan covering some 700 square kilometers - never reach the surface area of even small islands, for example, those smaller in size than Madeira or Singapore, the Isle of Man, or St. Lucia in the Antilles or Tonga in the South Pacific. And, over the course of the ages, such little islands have never sheltered beasts as colossal as the diplodocuses, the ceratosaurs or the iguanodons.

The largest creature which one could have hoped to find on the terraces of a *mesa* like Auyan Tepui would be a giant monitor lizard like that of the island of Komodo, which does not exceed 400 square kilometers in area. Moreover, the *tepui* in question would have to contain enough game to feed such lizards.

Besides, why do romantic minds become so enthralled about the case of the tepuis? After all, no matter how spectacular they are, they are no more 'islands in time' than ordinary islands in the seas which have been separated from their continents since ages long gone by. Furthermore, it is naive to believe that, in order to be isolated from the rest of the world except by air, and to be preserved from all destructive influences by competing forms, a fauna must necessarily be emprisoned on the summit of a *tepui* or on a remote island. A continent or an enormous island can equally well serve the purpose: Australia is 'the lost world' of the monotremes and of the majority of the marsupials; Madagascar, that of most of the lemurians; New Zealand, that of the last rhynchocephalous reptiles and of a formerly flourishing avian empire.

On the continents themselves, broad rivers, large lakes, mountain ranges, deserts of sand, dense forests, even very low temperatures - like a sort of wall of cold - for certain animals prove to be insurmountable obstacles. And, contrariwise, a particular habitat which is unhealthy for most creatures can be a sure refuge for the species which have succeeded in adapting to

it. A suffocating morsel of primary forest, the frozen tundra, a miry and noxious swamp, a very high mountain, a subterranean grotto, the desolation of sandy dunes - all are just as much 'lost worlds' - and all hold just as much promise for stunning new zoological discoveries.

The sum total of the major upsets which still shake up the field of geography alone is enough to stagger the imagination. Imagine, we continue to discover peaks and islands, even a hidden continent, we must still move mountain ranges about and add new tributaries to certain rivers, we see that some lakes have passed unnoticed, and we have missed the highest waterfalls in the world. Now, there is something which should make even the most obstinate disbeliever pause and reflect - and above all those who laugh in the face of people who claim to have encountered a medium-sized dinosaur, a little pterodactyl or a bipedal great ape, in short, creatures which are simple mice with respect to the unknown mountains which may have given birth to them.

Komodo dragon swimming, photograph by Sven Gillsater, Rapho.

89

THE 'TERRA INCOGNITA' OF ZOOLOGISTS

In truth, for the zoologist seriously devoted to the search for new discoveries, our entire planet can be termed an 'Unknown World'.

We are not even speaking here of ichthyologists, cetologists and other specialists who are seeking after the capture of still-unknown marine animals: the fauna of the oceans which, it must be recalled, cover about three-quarters of the globe, is so rich and that of the great depths so little known that, for them, all manner of hopes still are justified.

In the seas, every dip of a net can bring up a new species. Thus, for example, when a systematic programme of fishing was carried out in 1952 in waters as frequented as those of the Gulf of California, there were collected not less than some fifty unknown species of fish. It is not even necessary to go fishing oneself in order to make new discoveries. In fact, the ichthyologists who described these 'new' California fish discovered, after the fact, that several of their protégés were currently on sale in Mexican markets!

Let it be said in passing that it is not only in public markets that one is exposed to the risk of making sensational new findings. We know already that mammals and birds of new species and new genera have been discovered in collections in natural history museums (among others, the dwarf chimpanzee and the Congolese peacock) and even, on one occasion, in a zoo (Grevy's zebra). There was even found a human form which had been believed fossil (the pongoid Neanderthal) in an itinerant fair wagon.

With regard specifically to terrestrial and fresh-water fauna, one might think that our knowledge is much more complete than for the fauna of the sea, and our hopes of making new findings, therefore, rather more vain. This is far from being true, even outside of the 'blank spots' which are still scattered across the map of the world.

To be sure, in this domain it all depends on the size of the animals being sought. For the entomologist, for example, the entire world is his field of action. This is true also for the arachnologists, and *a fortiori* for those among them who are specialists in acarids. And, it is the same for the naturalists specialising in the study of other terrestrial or fresh-water invertebrates which, on the whole, are of small size. If they choose to take the trouble

to explore deep wells, grottoes or underground watercourses, they are even certain to turn up unexpected findings of great interest. Speleologists are constantly bringing up to the surface so many unknown organisms that a whole new branch of zoology - biospeleology - has been formed to record and study their discoveries. However, as the great English zoologist Maurice Burton writes, 'if we are content with small things, insects, spiders and the like, there is no need to search the inaccessible places; we may quite easily find new forms on our doorstep'.

Dr. Maurice Burton (1898-1992), photograph by B. Heuvelmans.

This being said, even if one considers only animals of a much more appreciable size, one would be grossly mistaken to imagine that they cannot pass unnoticed and that they are easy to spot. The current repertory of the majority of the macrofaunas (i.e., the whole of the groups of animals of medium to large size) cannot be kept up to date in any country in the world. Nowhere is it possible to establish unequivocally the complete listing of all species of mammals, birds, reptiles and amphibians. In each of these classes, there is always a very small percentage for which it is really not well known whether they still exist or not. A striking proof of this is given by the cataloguing of some important species which had been believed extinct for at least ten years, a period in which there was in any case more than enough time to carry out systematic and ongoing search efforts.

Among the animals contained in this listing, which is by no means limiting, of the 32 cases chosen there was only one amphibian and one reptile. On the other hand, there were about fifteen mammals and not less than fourteen birds, only one of which was incapable of flight. The presence in this listing of such a number of creatures which are so readily visible is really quite extraordinary. The eclipses recorded often lasted a number of decades, and sometimes more than a century, and in one case even 300 years. The average duration was about 75 years or, in other words, the time span of three generations of researchers!

If a known animal in a specific region and well-defined habitat can apparently disappear for such a long time, does one not have the right to suppose that some other ones are perhaps still hiding within the same country?

It goes without saying that city dwellers have little chance indeed of ever encountering wild animals other than sparrows and pigeons, rats and mice. However, we delude ourselves if we think that peasants and farmers, people who have remained close to the land, frequently see representatives of the wildlife of their countryside, and all the more so if these creatures are forest-dwelling or burrowing, or aquatic or nocturnal. How many country people in Western Europe, the most domesticated of all the continents - and especially including those in France, Belgium and Switzerland - can claim to have seen in the course of their entire lives carnivores as common and widespread as the fox, the otter, the badger, the polecat and the marten, to say nothing of the wildcat and the genet? Those who are most likely to have had such an experience clearly are the hunters and naturalists, for they make every effort to insure that this happens. But even these people rarely have the occasion to come upon these animals. So, draw your own conclusions.

On the occasion of a television program by Frédéric Mitterrand devoted to the wolf, the author, Bernard Heuvelmans, in an aside to his Swiss colleague, Robert Hainard, asked the latter how many times he had had the occasion to observe a genet in the wild. It must be emphasised here that Hainard, a field naturalist, an animal artist and author of the classic work *Mammifères Sauvages d'Europe* (Wild Mammals of Europe) is without doubt the one person in the world who has the most often studied these animals 'in their most secret retreats, in the heart of the wild', as was said of him in a short biography. He has passed more than sixty years without interruption in this field of activity, and his response was categoric: 'That never happened to me'.

If such a state of affairs is possible in Europe, whether it be in the valleys of Helvetia, on the soil of gentle France or of overpopulated Belgium, one cannot doubt that much more astonishing surprises await us in those regions of our planet which have remained far more wild.

In such regions, it is no longer little carnivores which pass unnoticed for decades; it is the big cats, the enormous apes and even the largest ungulates of all the terrestrial fauna. The 'lost world' is omnipresent.

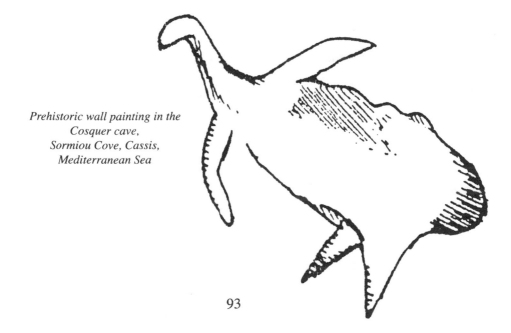

Prehistoric wall painting in the Cosquer cave, Sormiou Cove, Cassis, Mediterranean Sea

93

The author consulting Dr Oudemans' unpublished archives in 1959.

THE BIRTH AND EARLY
HISTORY OF CRYPTOZOOLOGY

'There are disconcerting facts affirmed by serious men who have witnessed them,
or who have learnt of them from men like themselves: to accept all or to deny
all seem to have equal disadvantages; and I venture to say that here, as with all
things out of the ordinary, not within the common rules, there is a course to be
steered between the credulous and the unbelievers.'

(Jean de la Bruyere, *Les Caracteres*, chap. XIV, 1689)

Almost until the end of the 18th century, zoology did not need
cryptozoology. A systematic search for animal species still unknown
was then quite superfluous. Since European travellers, particularly
from the 15th century on, had started to explore and conquer with insatiable
greed all of the 'lands beyond the horizon', netting and trapping animals
in all directions or, in fact, even just fishing and shooting them at random,
seemed amply rewarding for this purpose. *All* naturalists, burning with
curiosity and eager to discover anything new, were lending their ears to the
most vague rumours about animals, birds, reptiles or fish which appeared
still not to have been recorded. In a certain sense they were *all* consumed
with the spirit of cryptozoology, even though they did not feel the need to
build up a refined method to achieve their objectives within the shortest
possible period of time.

Above: Synbranchus.

Buffon said quite appropriately, 'The ancients, whose minds were less confined and whose thinking was broader, were less astonished than we are by facts which they could not explain: they saw more clearly nature as it is.'

Such intellect lasted into the Renaissance.

FROM OPEN-MINDEDNESS TO STUBBORN DISBELIEF

Never did the zoologists of the sixteenth and seventeenth centuries hesitate to admit into their catalogues or their general works any animal which was spoken of in the world, even if its mummified or pickled remains - shell, skin, skull or skeleton - were not present in the latest cabinets of curiosities or in the newly emerging museums of natural history. This, of course, had a disadvantage, although only a minor one: that of including in the manuals of zoology the descriptions and even the portraits of some little-known beings, often metamorphosed to such an extent by tradition that their original form was scarcely any longer recognisable. In fact, it ultimately turned out that these fabulous creatures (often called 'monsters' because of the extravagant and frightful traits attributed to them) were nothing other than perfectly well-known animals. They were unconsciously mythicised by our emotional thoughts, and thus were more or less distorted and 'romanticised' in order to fit the mould of a whole range of archetypes, reflecting our ambitions, our fears and our cravings, our prejudices and our internal conflicts.

The most fantastic of these creatures - like the unicorn and the satyr, the mermaid and the sea-serpent, the dragon and the basilisk, the phoenix and the roc-bird - figured in the majority of the great encyclopedias of the naturalists and geographers of the sixteenth and seventeenth centuries. But, a hundred years later, because these animals were imbued with mythical and sometimes supernatural characteristics, most of them had disappeared from the works of two of the most important zoologists of the eighteenth century, the Swedish Carl van Linné (1717-1778), who had endeavoured to classify nature in a hierarchical system, and the French Buffon (1707-1788), who for his part sought the causes and reasons for this diversity.

Not all of them, however, were eliminated. Thus, the *Homo troglodytes*, the hairy wild man of nocturnal behaviour, and the *Microcosmus marinus*, a tentacular beast so huge that it could be taken for an island, both still figured

Bishop Erik Ludvigsen Pontoppidan, Universitetets Zoologiske Museum, Copenhagen

in the *Systema naturae* of Linnaeus (at least in its earlier editions). And Buffon still believed that there were tigers in Africa, and even an abominable woodsman, a kidnapper of black girls, the Pongo, a truly unbelievable creature which we now classify under the name of *Gorilla gorilla*.

There is nothing really surprising in all of that. Scarcely fifty years earlier, an erudite and respected man, Bishop Erik Ludvigsen Pontoppidan (1698-1764), had devoted an important chapter of his Natural History of Norway to the three most prestigious stars of the Pantheon of sea 'monsters': the mermaid, the kraken and the sea-serpent. Scandinavia being a land rich in legends, these creatures could easily have been dismissed in other countries as fables born of the mists of the North. But, no scientist of that time thought of questioning their existence. Nevertheless, it must be said that even the most phantasmagorical creatures of ancient literature, art and folklore appeared quite prosaic as soon as equally unlikely animals were well and truly observed in distant lands, or their colossal remains were dug up from the bowels of the Earth.

It was thus that, in 1784, there appeared in Germany an anomyous booklet with a most significant title: *Beschreibung, ausführliche und accurate, nebst genauer Abbildung, einiger vorhin fabelhafter Geschöpfe, welche in der heutigen Naturgeschichte berühmter Schriftsteller gänzlich verändert und ins Light gestellt sind* (Detailed and Accurate Description, Together With Their Right Representation, of Some Formerly Fabulous Creatures, Which in the Present Natural History of Diverse Authors of Good Reputation Appear Completely Changed and Brought to Light). Some years later, in 1791, there was published in Italy a scientific dissertation of rather similar bent, by Dr Luigi Bossi, member of the Academy of Sciences and Humanities of Mantua, and written in the form of a letter addressed to Count Gian Rinaldo Carli: *Dei Basilischi, Dragoni et altri Animali Creduti Favolosi* (On Basilisks, Dragons and Other Animals Held to be Fabulous).

In short, it turned out that, in the light of the progress of science, animals could not be divided into two separate and distinct categories: those which were clearly real, on the one hand, and those which derived entirely from fables, in other words, the imagination, on the other. Beasts which had been believed legendary were encountered in nature and wound up stuffed and mounted in museums, or even alive in zoological parks. Gigantic bone fragments, dug up here and there, showed that there had existed - in former times in any case - animals which were more outlandish even than those which had been considered to be complete inventions of the imagination. And so zoologists did not hesitate to give to creatures of flesh and blood - sometimes the most prosaic of all - names (like dragon, basilisk, kraken, lamia, devil, sphinx, satyr, etc.) which had been considered as reserved for the bestiary of ancient mythology. In brief, fabulous animals were shorn of their fantastic traits, whereas real animals were decked out in the trappings of mythology. The frontier between the two began to fade little by little.

However, not for long. When observations of legendary monsters of the most classic types continued to be reported in the nineteenth century (sea-

Portion of Olaus Magnus's map of Scandinavia, 1539.

serpents along the Atlantic coast of North America, colossal octopuses off the coasts of Africa, and even gorgeous blond mermaids around the British Isles), certain nitpicking naturalists - armchair naturalists for the most part -.began to complain sharply and indignantly, with exaggerated force and sometimes violence, against what they called 'absurd fables', 'travellers' tales' (what comes from afar is a lie), 'fairy tales' and even 'chimeras generated by simple-minded or diseased imaginations'. The sea-serpents of the New World were stigmatised with the infamous term of 'Yankee Humbug'.

Such a brutal change in manner of expression by the scientists of the time calls for an explanation.

99

BARON CUVIER VERSUS PRESIDENT JEFFERSON

In 1795, the President of the United States, Thomas Jefferson (1743-1826), who was at least as well-versed in the natural sciences as in politics, had the occasion to examine some fossil bones which subsequently proved to be from a giant ground sloth, including, in particular, a truly enormous claw. It was for this reason that Jefferson gave to this animal the name of *Megalonyx*, which means 'great claw', but he considered it to be a giant feline 'more than three times as large as the lion'. On the basis of this concrete piece of evidence he concluded: 'If this animal then has once existed, it is probable on [the] general view of the movements of nature that he still exists'. He thus felt quite justified in declaring in a report published in 1799: 'In the present interior of our continent there is surely space and range enough for elephants and lions, if in that climate they could subsist; and for mammoths and megalonyxes who may subsist there. Our entire ignorance of the immense country to the West and North-West, and of its content, does not authorize us to say what it does not contain.' He said this for the reason that he had seen, on the steep rocky bank of a river, ancient carved representations of various animals, and among them 'a perfect figure of a lion' (not of a puma!), and also because the existence of such a wild beast was supported by Indian traditions 'considered as fables, but which have regained credit since the discovery of these bones'. This is a perfect example of the cryptozoological approach.

To be sure, Jefferson was mistaken about the identity of his Great Claw, but the credence which he gave the Indian legends was in no way diminished by this fact. These legends were doubtless based on the presence in North America, in the Pleistocene, of what is held to be a giant jaguar, *Panthera atrox*, not necessarily spotted, the remains of which were found subsequently and which surely was still alive some 10,000 years ago. In any case, no one at the time would have thought of dismissing or shrugging off the statements of President Jefferson.

Some fifteen years later things had changed considerably, when Georges Cuvier, the 'father of paleontology', declared in peremptory fashion in 1812 that there was 'little hope of discovering new species of large quadrupeds'. With regard to fabulous animals, he allowed himself some pointed sarcasm: 'we hope that no one will look for them seriously in nature: one could just as well search there for the animals of Daniel or the beast of the Apocalypse.

100

Let us not look there either for the mythological animals of the Persians, the offspring of an even more fertile imagination.'

We were then well into the nineteenth century, which has been termed, and not without good reason, 'stupid'. Henceforth, dogmatism, sectarianism, recourse to the position of authority in short, a rigid and recognised authoritarianism - was to permeate all of our knowledge under the pretext of rationalism and an ill-digested positivism. The decision of what was to be accepted or rejected by science - which was supposed to solve all of the problems of humanity, and was therefore deified - seemed to be inspired by some religious fanaticism. The time had come - or, more exactly, had *come again* from the depths of the Middle Ages - for pontification, academic excommunications with great pomp and ceremony, and scientific taboos.

In the realm of zoology, this trend manifested itself in the veritable intellectual dictatorship exercised by some scientists - who were nonetheless remarkable men - like Georges Cuvier (1769-1832) in France, Sir Richard Owen (1804-1892) in England and Rudolf Virchow (1821-1902) in Germany.

In the United States, i.e., in a society clearly without a past history of disparate peoples, the clan of the holders and dispensers of knowledge had not had the possibility of living for centuries through a cascade of scientific revolutions, and so of being convinced of the fragility and ephemeral character of all theories and all systems. There, the tendency to a narrow and rigid dogmatism was necessarily going to last for a much longer time than in the Old World: in fact, about a century longer. In zoology, it was above all incarnate in the person of the great paleontologist George Gaylord Simpson (1902-1984).

SCEPTICISM OPPOSED TO BOTH GULLIBILITY AND INCREDULITY

A few enlightened laymen, who could in a way be called the 'Prophets of Cryptozoology', even before the end of the eighteenth century had already sensed coming the crisis which science was soon to traverse. Shortly before his death, the English novelist Oliver Goldsmith (1728-1774) wrote in his essay *An History of the Earth and Animated Nature*: 'To believe all that has been said of the sea serpent, or the kraken, would be incredulity: to reject the possibility of their existence would be presumption'. And, in

1799, the English actor and dramatist Thomas Holcroft (1745-1809), after having heard of recent sightings of these very same sea monsters, wrote in a letter to one of his friends: 'who can affirm he can mark out the boundaries of possibility? Some mariners treat these tales as absolutely false and ridiculous; others seriously affirm them to be true; and I think it is a duty to collect evidence, and to remain on this question as on many others, in a certain degree of skepticism.'

So there, the word has been spoken. As in everything involving science, 'scepticism' is the key word here, for it sums up the true spirit of scientific research. And it is precisely because this term carries so much weight that, deliberately or unconsciously, it is almost always used in an improper, false or incorrect sense by most people, and particularly by those who should be branded with the epithet of 'systematic debunkers'. When these people are confronted with new ideas or new facts - whether they be unusual, astonishing, disturbing or embarrassing - they content themselves with shrugging their shoulders and asserting that, being *sceptics*, they cannot believe in such stories. But, in reality, they are not in any way sceptics; they are simply incredulous, which is a quite different, and even contrary, thing. The essence of scepticism is, in fact, doubt. To be a true sceptic, as was so well said by Anatole France (1844-1924) , who was one, one must 'doubt everything, even whether one's doubts are well-founded'. One neither accepts nor rejects anything out of hand. One expresses no opinion before having been fully informed, and before having examined thoroughly and with great care *all* possible proofs.

Pierre Bayle (1647-1706), the author of the *Dictionnaire Historique et Critique* (Historical and Critical Dictionary), whose spirit seems to have inspired that of the whole following century, the Century of Enlightenment, said quite justly: 'To believe nothing or to believe everything are extremes, neither of which is worth anything'. It goes without saying that there is a happy medium to be found between what Father Jacques d'Autun as early as 1671 termed 'scientific disbelief and ignorant gullibility'.

When all is said and done, cryptozoology is founded on the most elementary of common sense.

THE GREATEST NAMES OF SCIENCE HAVE SUPPORTED THE SPIRIT OF CRYPTOZOOLOGY

Men of science with open minds never yielded - need it be said? - to the ukases of the scientific dictators of the nineteenth century. They continued to show the same interest in all undesirable 'monsters', and to collect zealously the slightest scraps of information about them. The fact is that the legendary sea-serpent did not appear only in the tabloid gazettes, nor exclusively during the summer silly season, as popular belief would have it, although without the slightest justification. It was mentioned regularly in the scientific literature of the entire world. To be persuaded of this, it is enough to glance through the indexes of some of the most respected scientific publications of the last century.

Edward Newman (1801-1876), the editor of the *Zoologist*, of London, in the preface to the first volume of his journal (1847), summed up very well what was, or at least ought to have been, at that time the attitude of true sceptics: 'the communications and quotations about "the *Sea Serpent*" are well worthy of attentive perusal: it is impossible to suppose all the records bearing this title to be fabricated for the purpose of deception. A natural phenomenon of some kind has been witnessed: let us seek a satisfactory solution rather than terminate enquiry by the shafts of ridicule [...] surely it is not requiring too much to solicit a suspension of judgment on the question of whether a monster may exist in the sea which does not adorn our collections.'

Among the earliest supporters of what was going to become cryptozoology, the greatest names of science could be mentioned. In fact, I did this in my work *Le Grand Serpent-de-Mer* (The Great Sea-Serpent) (1965) for the particular problem in question, mentioning those eminent men of science who were particularly interested in the great sea-serpent.

However, the eyes of curious naturalists were not focussed on these sea monsters alone during the last century. In his imposing work of twelve volumes which he devoted to his *Voyage aux Regions Equinoxiales du Nouveau-Monde* (Travels Through the Equinoctial Regions of the New World) - undertaken from 1799 to 1804 in company with the French scientist Aime Bonpland (1773-1859) - Alexander von Humboldt (1769-1859), the 'father of physical geography', expressed certain doubts regarding the

existence in South America of a great anthropoid ape, concerning which many rumours were then circulating. This being said, he nevertheless refused categorically to consider them to be simple fables: 'In treating them with disdain, the traces of a discovery may often be lost, in natural philosophy as well as in zoology [...] Travelers who may hereafter visit the missions of the Orinoco will do well to follow up our researches on the *salvaje* or *great devil* of the woods; and examine whether it be some unknown species of bear, or some very rare monkey [...] which may have given rise to such singular tales.'

What encouragement from on high for cryptozoology!

Mastodonsaurus by Zdenek Burian.

104

IN GREAT BRITAIN, SURVIVING FOSSILS COME INTO FASHION

In another field, and in a little different vein, the freethinkers of science soon also raised opposition to a dogma of Cuvier other than that which condemned explorer-zoologists to despair: that one according to which the animals of 'prehistory' belonged to ages long past, had never been contemporaries of Man, and thus could certainly not be surviving today and so could not possibly explain certain mysterious creatures of our times.

One of the first to envisage the possibility of such survivals was the French naturalist-explorer Jean-Baptiste Bor de Saint-Vincent (1780-1846), who is considered to be one of the pioneers of the popularisation of the natural sciences. In the article 'Lost Animals' in the *Encyclopedie Moderne* (Modern Encyclopedia), published from 1822 to 1830, he did not hesitate to write that, if Indian traditions are to be believed, the mastodon is still living, probably around the upper St Lawrence valley, in Canada. Furthermore, he confirmed the suggestion of President Jefferson regarding the survival of the megalonyx, which had been identified in the meantime as a ground sloth of very great size.

However, it was above all in Great Britain that men of science began to put forth a multiplicity of similar hypotheses. In 1844, for example, the Scottish geologist Hugh Falconer (1808-1865), after having discovered in India the fossil remains of a truly gigantic sea turtle (which he had named *Collossochelys atlas*), raised the question whether, by chance, this mesozoic reptile might not have survived into historic times, thus giving rise to the Hindu legend about the giant turtle which supports the elephant on which the Earth rests.

A few years later, in 1848, another British naturalist, of Flemish origin, Colonel Charles Hamilton Smith (1776-1859), also surmised from American Indian legends that the redskins had actually known living mastodons and giant ground sloths, which moreover was later to be confirmed by archeological and paleontological findings. However, it should further be said in passing that he also supported the theory according to which the unicorn was a real animal, still prowling the depths of darkest Africa. This was ultimately to lead to the discovery of the northern race of the white rhinoceros (*Ceratotherium simum cottoni*) in 1900.

105

After C. H. Smith, a third British naturalist, the Englishman Charles Carter Blake, author of a manual of zoology for students which had nevertheless remained obscure, was in his turn impressed by the traditions of the Indians of Brazil regarding a great anthropoid ape in their region, the Caypore. He wrote in 1863, 'No such ape exists in the present day; but, in the post-Pliocene in Brazil, remains have been preserved of an extinct ape (*Protopithecus antiquus*) four feet [1.20 m] high, which might possibly have lived down to the human period, and formed the subject of the tradition'.

In each of these cases, scientists were not at all concerned with the present existence of still unknown animals, but rather with the survival of animals, already known from their fossil remains, into an age when they were believed no longer to exist, or perhaps even into the present time. This being said, as these animals were in a certain way 'hidden' from the eyes of science, searching for them revealed a true cryptozoological turn of mind.

The ideas of Falconer, Smith and Blake were to be taken up with enthusiasm by Edward Burnett Tylor (1832-1917), the 'father of ethnology' (called 'social anthropology' in the Anglo-Saxon countries). In his epoch-making work *Researches into the Early History of Mankind* (1865), he emphasised the point that popular traditions have preserved up to modern times the memory of certain 'prehistoric' animals claimed to have disappeared before the arrival of *Homo sapiens* (we would speak today more precisely of animals of the Lower Pleistocene).

This particular aspect of cryptozoological research was admirably defined in 1886 by the English geologist Charles Gould, the only son of the great ornithologist and portrait artist of birds, John Gould (1804-1881). Here is what he wrote in his outstanding book *Mythical Monsters*: 'I have but little hesitation in gravely proposing to submit that many of the so-called mythical animals, which through long ages and in all nations have been the fertile subjects of fiction and fable, come legitimately within the scope of plain, matter-of-fact natural history, and that they may be considered not as the outcome of exuberant fancy, but as creatures which really once existed, and of which, unfortunately, only imperfect and inaccurate descriptions have filtered down to us, probably much refracted, through the mists of time.'

Let us say in this regard that the Irish naturalist Valentine Ball (1843-1895) had just published in 1884 two articles in which he had endeavoured to identify certain legendary creatures of Ancient Greece, such as the pygmies, the martikhora and the griffin.

A new wave was breaking over British zoology.

FIRST STEPS IN THE STUDY OF HIDDEN ANIMALS IN GERMANY AND IN FRANCE

In the last century there was also great interest in German scientific circles in the possible survival of animal species considered extinct, as well as in the possibility of exciting discoveries of new forms of animals of medium to large size.

In 1841, Heinrich Rathke (1793-1860) , one of the pioneers of the science of animal development, thus declared that there could be no doubt about the existence of the Norwegian 'seaserpent'. In 1858, in the journal *Die Natur*, which he published in Jena, Karl Müller (1818-1899) devoted a long series of articles to the efforts at identification of various 'mythical animals'. And, in 1877, the great herald of evolutionism, Fritz Müller (1821-1897), discussed at length not only the traditional 'sea-serpent' but also a supposed 'fresh-water sea-serpent', if one may call it that, which was supposed to frequent the Amazon and Orinoco basins, the Minhocao. According to his expert opinion, this aquatic monster could well be 'a gigantic fish related to the *Lepidosiren* and the *Ceratodus*', the enigmatic lungfishes.

In fact, this suggestion had already been advanced some fifty years earlier, in 1830, by the French explorer and botanist Augustin François Cesar Prouvençal de Saint-Hilaire, better known under the name of Auguste de Saint-Hilaire (1779-1853), who thus proved to be one of the very first contributors to cryptozoology.

Even in France, some interest had been shown in fabulous beasts, which was rather unexpected in the country which prided itself on being the cradle of rationalism, and where it was thought that expressing disbelief was a sign of reason. Baron Henri Marie Ducrotay de Blainville (1778-1850), professor of zoology and comparative anatomy at the Museum of Natural History of Paris, in 1818 was among the first supporters of the sea-serpent.

In 1826, a distinguished historian and jurist, Eusebe Baconniere de Salverte (1771-1839), whom the celebrated physicist François Arago (1786-1853); considered as 'one of the most erudite men' of his time, published a long and very well documented study entitled *Des Dragons et des Serpents Monstrueux qui Figurent dans un Grand Nombre de Recits Fabuleux ou Historiques* (Dragons and Monster Serpents which Figure in a Great Number of Fabulous and Historic Accounts). And, under the pseudonym of A. Frédol, the emininent malacologist Alfred Horace Bénédict Moquin-Tandon (1804-1863) devoted a portion of his posthumous book *Le Monde de la Mer* (The World of the Sea) (1865) to the story, still controversial at the time, of the discovery of gigantic cephalopods.

THE FIRST WAVE OF POPULAR CRYPTOZOOLOGY

It would certainly be unfair not to mention at once all of the popular science writers of the nineteenth century, who contributed so strongly to making the general reading public conscious of the problem posed by supposed 'monsters' of all sorts.

As early as 1818, in Great Britain, an anonymous author who signed himself 'W.', had written two quite exhaustive articles on the kraken and the sea-serpent, which were published in *Blackwood's Magazine* of Edinburgh. They were even commented on in a following number by a certain 'W.B.', who was, to be sure, none other than the editor-in-chief himself, William Blackwood (1776-1834). The 'essential marrow' of the material was prepared for the *Retrospective Review*, of London, but it never appeared there. However, its French translation was published in the *Revue Britannique* of Paris in 1835. That same year it was adapted in German by H. M. Malthen in his *Bibliothek der neusten Weltkunde* (Library of Recent World Science). A rehash of the version published in the *Revue Britannique* had been concocted in the meantime by Jules François Lecomte (1814-1864) for the pages of *Musée des Familles* of Paris, where it appeared in fragments in 1836 and 1837.

In 1849, an American poet, Eugene Batchelder (1822-1878), carried enthusiasm to the point of composing, under the pseudonym of 'Wave', a *Romance of the Sea-Serpent or Ichthyosaurus*, backed up by ancient and modern writings on the problem, as well as by letters emanating from men of science and from captains of the merchant navy.

Then, in 1863, a young member of the Geographical and Anthropological Societies of Great Britain, William Winwood Reade (1838-1875) published, under the title of *Savage Africa*, the account of his recent travels across the dark continent. His work contained not only first-hand information on the gorilla, which had just been described, and on the stories of unicorns, men with tails, etc., but also some brilliant thoughts which could well serve as fundamental principles for cryptozoology. For example: 'It must be laid down as a certain principle, that man can originate nothing; that lies are always truths embellished, distorted, or turned inside out. There are other facts beside those which lie on the surface, and it is the duty of the traveler and historian to sift and wash the gold grains of truth from the dirt of fable.' He also wrote, 'Incredulity has now become so vulgar a folly, that one is almost tempted, out of simple hatred for a fashion, to run to the opposite extreme. However, I shall content myself with citing evidence reflecting certain unknown, fabulous, and monstrous animals of Africa, without committing myself to an opinion one way or the other, preserving only my convinction that there is always a basis of truth to the most fantastic fables, and that, by rejecting without inquiry that which appears incredible, one throws away ore in which others might have found a jewel. A traveler should believe nothing, for he will find himself so often deceived; and he should disbelieve nothing, for he will see so many wonderful things; he should doubt; he should investigate; and then he may perhaps discover.'

In 1879, the English astronomer Richard Anthony Proctor (1837-1888) followed these directives to the letter in writing chapters on the sea-serpent and the mermaid for his *Pleasant Ways of Science*. Additionally, the celebrated popular naturalist of the Victorian era, Philip Henry Gosse, (1810-1888), went still further into the question of animals which remain to be discovered, in *The Romance of Natural History* (1860). He dealt not only with the case of the kraken and that of the sea-serpent, which he called 'the Great Unknown', but also with those of terrestrial creatures not yet described, like the African unicorn and the anthropoid ape of South America, which intrigues us still today.

Numerous tales were circulating around the world about such 'ape-men', as well as about 'man-apes', popularised by the fashion of the Darwinian revolution. To be sure, there were accounts of the ancient gorillas by the Carthaginian admiral Hannon; of the Soko discovered by Livingstone in the east of the Congo; of a fearful hairy creature in China with the coaxing

laughter of a young girl, the Fesse (in reality Feifei); and of the Susumete of Honduras (more exactly, Sisimite, from the name in the Nahuatl language, Tsitsimitl). All of these tales finally led the writer Philip Stewart Robinson (1847-1902), in his *Noah's Ark* (1882), to say of Professor Thomas Huxley, who was then nicknamed 'the bulldog of Darwin': 'If he were only to travel tomorrow into an unknown land, I am not at all sure that he would not ultimately emerge from some primeval forest hand-in-hand with the "missing link".'

THE FASHION OF FABULOUS BEASTS

In France, in the meantime, the folklore author A. Leroux de Lincy (1806-1869) had felt called upon to devote one of the introductory chapters of his *Livre des Légendes* (Book of Legends) (1836) to folktales relating to animals and, inspired in this regard by the article in the *Revue Britannique* of June 1835, which he had just read, he examined with interest 'the apocryphal animals, all of these monsters in strange, supernatural forms, the existence of which was considered as beyond doubt in the first sixteen centuries of our era'. He felt compelled to note that dragons, flying snakes, griffins and leviathans had well and truly existed in former times, as was then proved by bones found in the earth, and which scientists such as Cuvier used to reconstruct the anatomy of similar monsters.

Thirty years later, the popular science writer Armand Landrin (born in 1844) devoted an entire book to various *Monstres Marins* (Sea Monsters) (1867) in the very popular *Bibliothèque des Merveilles* (Library of Marvels), directed by Edouard Charton (1807-1889), creator-founder of the *Tour du Monde* (Around the World), the celebrated *Nouveau Journal des Voyages* (New Journal of Travel).

In Germany, another popular writer, Emil Arnold Budde (1842-1921) dealt at length with the problem of the sea-serpent and its freshwater counterpart, the Minhocão, in his *Naturwissenschaftliche Plaudereien* (Informal Chats about Natural Science) (1891).

In England, the reality of fabulous animals experienced an unprecedented wave of popularity. The future editor of *Land and Water*, Francis Trevelyan Buckland, called Frank (1826-1880), never missed an occasion in the four

Facing page: Sea serpent with mammalian traits, drawn by John Rike for the magazine True, following the directions of Ivan T. Sanderson (1949).

110

successive volumes of his *Curiosities of Natural History* (1857-1865) to speak of the kraken, the Australian bunyip, furry snakes or of surviving moas. A distinguished polygraph, Frederic Edward Hulme (1841-1909), devoted an entire work, *Natural History Lore and Legend* (1895), to the fabulous beasts of antiquity and the Middle Ages and to 'their varying degrees of reliability'. And, the writer John Timbs (1801-1875), in his *Eccentricities of the Animal Creation* (1896), likewise raised questions concerning the reality of two particular 'monsters', the mermaid and the unicorn.

It is appropriate to emphasize that, in the last century, the majority of these open-minded naturalist and popular-science writers were simply reporting what had been said about all mysterious animals, and limited themselves to conclusions about the more or less strong likelihood of their real existence. They did not concern themselves with setting forth a personal opinion as to their identity. If their accomplishments were to be given a scientific name, one should speak of 'cryptozoography' (description of hidden animals), rather than 'cryptozoology'.

Chelodina longicollis of Australia.

TOWARDS A MORE SCIENTIFIC EVALUATION OF ALLEGED 'MONSTERS'

Only a handful of scientists took the risk, in the last century, of investigating how one or another of the hidden animals should be classified, i.e., of determining in what zoological category it should be placed. Certain ones, for example, at first accepted blindly that 'sea-serpents' could be considered merely as outsize snakes. Nevertheless, when the first fossil skeletons of the enormous reptiles of the Secondary were unearthed and their external

Mososaur by Zdenek Burian.

appearance was reconstituted, most of the more adventurous zoologists at once related the so-called 'sea-serpents' to marine representatives of this class, notably, the mososaur, which had the appearance of a gigantic snake with two pairs of flippers, and the plesiosaur, which rather resembled a giant long-necked sea turtle. This latter hypothesis, which subsequently became a classic, was supported in 1833 by the English geologist Robert Bakewell (1768-1843) in his *Introduction to Geology*. Henceforth, one would be obliged when speaking of sea-serpents to emphasise that they were not serpents at all, any more than sea-elephants are elephants or titmice are mice.

113

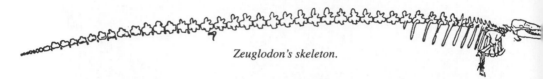

Zeuglodon's skeleton.

In the month of January 1835, Thomas Thompson (1788-1874), one of the vice presidents of the *Hull Literary and Philosophical Society*, presented before this learned English society a paper entitled *An Attempt to ascertain the Animals designated in the Scriptures by the Names Leviathan and Behemoth*. At the completion of an extremely well-documented study, Thompson did not hesitate to conclude that the former animal ought to be identified as the carnivorous megalosaur of William Buckland and of Cuvier, and the latter as the herbivorous iguanodon of the same Cuvier and of Gideon Mantell. In brief, he recognised each of these animals to be dinosaurs, rather distant from one another.

In fact, these first efforts at identification of fabulous beasts were more in the nature of a guessing game than zoological diagnosis, for they were based solely - and hastily - on the rather vague profiles of the creatures in question.

An English naturalist, the Reverend John George Wood (1827-1899), set himself apart from all others when, on the occasion of a visit to the United States, he had an article on the sea-serpent published in the *Atlantic Monthly*, of Boston, in June 1884, which was without a doubt the most penetrating ever written on the subject up to that time. In the light of a remarkable study in comparative anatomy, Wood concluded that, on the basis of its movements and behaviour, as well as its morphology, the 'sea-serpent', so often observed off the coast of New England, 'belongs not to the saurians, but to a cetacean animal, which, if not an actual zeuglodon, has many affinities with that creature'.

This being said, at the beginning of the nineteenth century, and thus well before the Reverend Wood, two nonconformist naturalists had already endeavoured to deal more seriously with the problem of the most popular sea monsters, and to study it with all of the rigour of the scientific method: Pierre Denys de Montfort (1764-1820) and Constantin Samuel Rafinesque (1783-1840). They were both French-born, but the latter was going one day to become an American citizen.

Zeuglodon by Zdenek Burian.

The colossal octopus of Pierre Denys de Montfort,
as shown in the Suites à Buffon, 1801-1802.

Denys de Montfort worked at the Museum of Natural History of Paris, where he was charged with writing the volumes relating to mollusks in the additions to the work of Buffon, edited by Charles Nicolas Sigisbert Sonnini de Manoncourt (1751-1821). In the second of the three volumes planned, and which was published in the year X of the revolutionary calendar (namely, in 1801-02), Denys described at length two cephalopods of truly gigantic proportions. He called the first 'the kraken octopus', referring back to Bishop Pontoppidan (we now know that it is not a true octopus but a super-giant squid, known under the name of *Architeuthis*). The second was 'the colossal octopus', which was described in 1897 under the name of *octopus giganteus* by Addison Emery Verril (1839-1926), although this latter repudiated it shortly thereafter as just 'a piece of whale blubber'. As we shall learn later on, the *Octopus giganteus* was to be rehabilitated only in 1971.

116

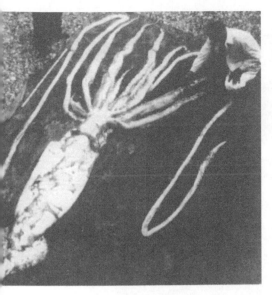

e Architeuthis found in Ranheim, Norway, 2 October, 1954 (photo Dr. Erling Silversten).

The Architeuthis stranded in Nigg Bay, Scotland, 1949 (photo Aberdeen-Bon-Accord).

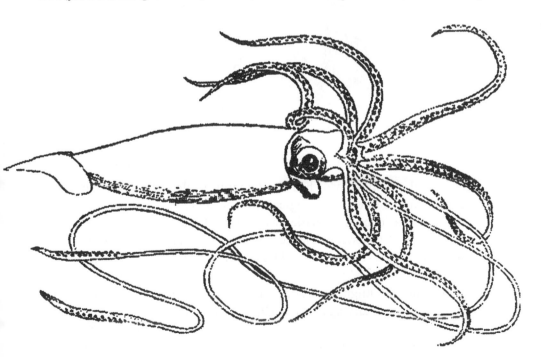

The first official portrait of Architeuthis, after Professor Addison E. Verrill, 1875.

Constantin Samuel Rafinesque (1783-1840), American Museum of Natural History.

The second naturalist who had the temerity to describe a sea 'monster' according to the rules of zoological nomenclature was Rafinesque, whom historians of science have hailed as 'the most remarkable man to appear in the annals of American science', according to the very words of David Starr Jordan (1851-1931). Donald Culross Peattie (1898-1964) said of him: 'Among all the naturalists who have ever worked on the American continent the only one who might clearly be called a Titan'. In any case, he was the very first, in 1817 and again in 1819, to publish a *Dissertation on Water-Snakes, Sea-Snakes and Sea-Serpents*, of extreme scientific rigour. After having studied all of the information available at the time on this subject, he concluded that, by reason of their anatomical characteristics, there existed

four different types of sea-serpent, two of which were nothing other than simple sea snakes of exceptional size, the two others being fish, one of which appeared to be related to synbranchid eels. From these somewhat hasty diagnoses, an essential element which must be retained above all is the fact that Rafinesque gave to the sea-serpent of New England the Linnaean scientific name of *Megophias monstrosus*.

Denys de Montfort and Rafinesque were so criticised, ridiculed, insulted and ostracised by their colleagues because of their writings that both finished by dying in appalling misery, the former, done in by alcohol, in a Paris gutter, and the latter, wasted by a generalised cancer, in a hovel in Philadelphia. Both were 56 years of age. These two ill-fated geniuses deserve to be considered as the true forerunners of the science of hidden animals.

THE BIRTH OF CRYPTOZOOLOGY

It is beyond question that every new science and every new original scientific discipline are the most clearly defined by their methods as much as by their objectives. Just as writing entertaining stories about animals does not make a writer a zoologist, so reporting encounters with unidentified animals of all sorts does not make a reporter a cryptozoologist. So, this is the reason why the publication in 1892 of the work *The Great Sea-Serpent* by the Dutch zoologist Antoon Cornelis Oudemans (1858-1943) constitutes the most decisive step forward in the building of cryptozoology.

Doubtless stimulated by the brilliant presentation of the Reverend J. G. Wood, Oudemans insisted on having more precise and more complete information, not only on the external anatomy of the sea-serpent, but also on its physiology, i.e., the various functions of its organism, and on its ethology, that is, its behaviour, as well as on its geographic distribution, in order to be able to describe its position in the system of nature with the maximum possible precision. In order to achieve this, he made use primarily of the statistical system used by the Austrian physicist Ernst Florens Chladni (1754-1827) to prove the extraterrestrial origin of meteorites. This had been a revolutionary discovery at the time, for these stones, which for centuries had been said to fall from the sky, had stuck in the craws of most men of science, rather like the 'flying saucers' of our days.

What Chladni had done for celestial stones, Oudemans did for sea-serpents. First, with great care and patience he collected from books, magazines and newspapers as many reports of sightings as possible (187 in all), and he then eliminated the obvious hoaxes and cases of possible mistaken identity. Following this, he extracted from the remaining ones all of the details available, which he then compared systematically one by one, to see if they matched or not. Finally, he drew a sort of identikit picture of the animal which had been so often observed. In this way it turned out that the creature appeared to be a species of long-necked and long-tailed seal or sea-lion, which Oudemans described scientifically under the name of *Megophias megophias*, the generic name having been taken, with all due credit, from Rafinesque.

Cryptozoology was born at last. The great work of Oudemans can be considered as the true starting point of the new discipline, just as *Recherches sur les Ossemens Fossiles* (Research on Fossil Bones) of Cuvier marks the beginning of paleontology as a science.

The views of the Dutch scientist were soon adopted here and there in Europe by some reputable zoologists, like the Austrian oceanographer Emil von Marenzeller (1845-1918), as early as 1894, and then later on by his Franco-Rumanian colleague Emile G. Racovitza (1868-1947), of the marine biology laboratory of Banyuls-sur-mer, and by the leading French mammalogist and professor at the National Museum of Natural History, Edouard Louis Trouessart (1842-1927). These latter two espoused fully the thesis of Oudemans in 1903, after numerous sightings of sea-serpents had been made, from 1893 onward, off the coast of Indochina by officers and crewmen of numerous ships of the French navy. Ten years later, in 1913, Professor Edmond Perrier (1844-1921), director of the Museum of Paris, was to declare, with notable respect, that the book of Oudemans had been 'a veritable act of courage'.

However, it turned out that, in the accomplishment of his research, the Dutch zoologist had been slightly preceded (this fact was discovered only recently) by a great Swedish ethnologist, Gunnar Olof Hyltén-Cavallius (1818-1889). From 1883 on, in fact, this latter had undertaken a broad and systematic investigation of the animals known locally as Drake (dragon) or Lindorm (wyvern) and which were reported to be seen in some lakes of the Kronoberg region, notably the Yen, the Rottnen and the Helgasjon, in

the southern part of the country. Upon completion of this investigation, in 1885, he published a pamphlet entitled *Om Draken eller Lindormen* (On Dragons or Wyverns), which contained a total of 48 detailed eyewitness reports, but in which he hazarded no opinion as to the zoological identity of the animals in question. In this respect his work clearly could not rival that of Oudemans.

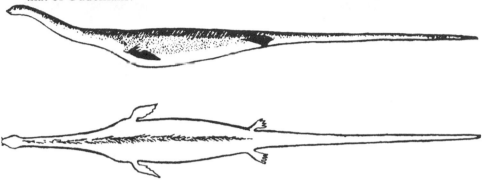

Megophias, side and top views, after Oudemans.

In 1899, following the appearance of the work of the latter, another Swedish researcher, the college teacher Peter Olsson (1838-1923), made his own investigation on 'monsters', but confined it to those of Lake Storsjö, in the Jämtland, this time in the northern part of the country. From the study of 22 credible eyewitness reports which he collected, he concluded, in his small book *Storsjöodjuret, Framställning af Fakta och Utredning* (The Monster of the Storsjö, an Exposé of the Facts and Solution), that the creature was an unknown pinniped, similar to the *Megophias* of Oudemans, but which for its part lived in fresh water. This was, in fact, the very first attempt to deal scientifically with the case of an unidentified lake animal, well before the famous affair of the 'monster' of Loch Ness burst upon the scene, in 1933.

THE FIRST MAJOR SUCCESSES OF THE CRYPTOZOOLOGICAL APPROACH

It is often asserted nowadays that cryptozoologists are losing their time and wasting their energy on a useless wild goose chase, since no unknown animal has ever been found by them. Such blather betrays a total ignorance of the history of zoology. Well before a method of research had been proposed by Oudemans, and in any event before such a method had been applied systematically, some dazzling victories had already been achieved empirically.

Dr Antoon Cornelis Oudemans (1858-1943).

The first marked success of the cryptozoological approach was the discovery of the mountain tapir by Dr François Désiré Roulin (1796-1874). Here is what this naturalist and traveller, who was also assistant librarian of the Institut de France, was to disclose on 9 February 1829 before the Academy of Sciences: 'Long before becoming acquainted with this second species of American tapir, I had been led to suspect its existence, less, I must admit, on the basis of generalities than on the strength of reports by old Spanish historians. Several of these writers, in fact, attributed to the tapir a thick coat of blackish-brown fur, a characteristic which does not match at all with the tapir of modern naturalists, that is, the one which I had seen myself in the plains and the broad low valleys which lie just above sea level.'

This shows quite clearly that the very first task of the cryptozoologist is intensive bibliographical research: the patient and painstaking perusal, sometimes tedious but always time-consuming, of accounts of travels and explorations or simple reports of hunting or fishing, as well as of old or exotic chronicles and ancient or foreign works on natural history.

When Roulin proposed to call the new species *Tapirus pinchaque*, it was because the latter word was the name of a fabulous animal of Indian folklore in the Popayan region of Columbia (in actual fact, *pinchaque* or *panchique*, which denotes a sort of 'hairy ghost' or 'werewolf'). The French physician was convinced that the belief in this much-feared monster was based on a rare and little-known tapir of the high Andes. He even went so far as to suggest that the white-backed tapir of India was likewise the basis for the legend of the griffin of Ancient Greece. To this, he added the following appropriate commentary, which emphasises the second important task of the cryptozoologist, that is, the careful and painstaking analysis of myths, legends and folklore of all the peoples of the world, in order to uncover 'monsters' for possible subsequent demythification: 'Having spoken the words *fabulous animals*, I feel the need to justify myself for having occupied the Academy with considerations so foreign to those with which it customarily is concerned. However, it is nonetheless true that this order of research cannot remain foreign to the natural sciences. It is impossible to trace the history of animals through ancient times without having at each moment to separate the real facts from the fables which surround them. If we do not have to discard even more of these, it is because this work of elimination has been going on, without our having been aware of it, for thousands of years.'

A second notable success of cryptozoology, and much more sensational in nature, given the circumstances, was the description in 1856 of the super-giant squid by the Danish researcher Johan Japetus Steenstrup (1813-1897). This scientist was the first to establish, on the basis of historic documents alone, that the kraken, the tentacular bogeyman of Scandinavian folklore - a monster even much more incredible than the sea-serpent itself - was nothing other than a giant cephalopod. According to the opinion which he first expressed in 1849, it was not an octopus, as Denys de Montfort had supposed, but an outsize squid, like the two huge specimens which, according to old archives, had stranded on the coast of Iceland in 1639 and 1790. The Danish naturalist then suggested, in 1855, that another legendary marine monster, the *Monachus marinus* (sea monk) of Rondelet, was likewise based on a huge squid which had been taken in a net in 1550 on the Swedish coast. And, finally, when he laid hands on the enormous beak of a specimen found dead on a beach of Jutland, in Denmark itself, and then on another beak, which was slightly different, from a specimen discovered floating in the sea between Bermuda and Carolina, he felt justified in describing the Scandinavian species under the name of *Architeuthis monachus* and that of the Atlantic under the name of *Architeuthis dux*.

FROM VICTORY TO VICTORY

The third capital success of cryptozoology caused a sensation throughout the entire world at the beginning of the 1900s.

In 1891, the celebrated Welsh journalist and explorer, Henry Morton Stanley (1841-1904) had reported casually in one of his books that the Wambutti pygmies 'knew a donkey and called it *atti*. They say that they sometimes catch them in pits.' These few lines pushed Sir Harry Hamilton Johnston (1858-1927), then governor of Uganda, to organise a patient but persistent search, which was going to lead after some ten years to the description of an animal as improbable as a forest zebra: zebras, like all horses, range across the savannah; after all, they have no business in the forest! However, this did not stop it from being described in 1909 under the name of *Equus johnstoni* by the English zoologist Philip Lutley Sclater (1829-1913), on the basis of a few pieces of striped skin. It was only after the study of a complete carcass, and especially of two skulls, that another English zoologist, Sir Edwin Lankester (1847-1929) was obliged to conclude that the *atti*, better known locally by the name of *o-api* or okapi, was not a zebra

at all, but rather a sort of short-necked giraffe, with stripes only on its rump. It appeared to be more or less related to a Miocene fossil from Greece, the *Helladotherium*. Well deserving of a genus of its own, the surviving proto-giraffe thus became *Okapia johnstoni*. The announcement of its discovery not only caused a stir into the very boondocks of zoology, it made page one of journals around the globe.

The fourth major victory of cryptozoology has taken much longer to be accepted - more than a century - and there are still those who have never taken the trouble to learn about the matter, and who even continue to raise doubts about it. This was the revelation of the survival into our times of Neanderthal Man (right).

In fact, a vague inkling of this had been provided in 1863-64 by the Swedish ethnologist Hyltén-Cavallius, who did not content himself in cryptozoology with going off to hunt the lake monsters of his country. Earlier, in his ethnological essay on Sweden, *Wärend och Wirdarne, ett Försök i Svensk Ethnologi* (The Region of Värend and its People, an Investigation in Swedish Ethnology), he had advanced the audacious hypothesis that trolls, these hairy wild men of Scandinavian folklore, might represent the recollection in the popular mind of the existence, in the Stone Age, of a primitive race very different from ours. It must be emphasised that, at that time, the skull of a Neanderthal Man had just been unearthed, in 1856, and that this archaic human form was to be generally recognised as a distinct and genuinely prehistoric species only in 1886, following the discovery at Spy, in Belgium, of skeletons dating unequivocally from the Pleistocene. This imparts a truly prophetic character to the opinion expressed by Hyltén-Cavallius.

To be sure, since his work had been published only in the Swedish language, the hypothesis set forth had little impact. In any event, precious few anthropologists and ethnologists were prepared to accept readily that traditions could be handed down through several tens of thousands of years. To accept the idea that their contemporaries could recall men of

the Paleolithic era, it was necessary to wait until the Soviet historian Boris Feodorovich Porshnev (1905-1972) set forth, in 1963, the hypothesis of the survival of Neanderthals into historic times and even until our days. If, as recounted in ancient chronicles, hairy wild men were still being encountered in Europe in the eighteenth century, and notably on the Swedish island of Öland, then it becomes quite probable that such creatures were the origin of the belief in trolls. Hyltén-Cavallius had perceived clearly, and an extraordinary scientific discovery was going to justify him fully.

In 1968, in fact, while I was on a visit to the United States, I, in company with my colleague and friend Ivan T. Sanderson, had the occasion to examine at length the deep-frozen cadaver of a man who was as hairy as an anthropoid ape, and who exhibited numerous anatomical traits unknown in *Homo sapiens*. This specimen had been imported illegally into the country from Vietnam, where it had been killed. The study of it disclosed that it was a specialised form of Neanderthal, which the Franco-Belgian zoologist described under the name of *Homo pongoides* (sp. Seu sub sp. nov). This discovery proved in incontrovertible fashion the legitimacy of the hypothesis of Porshnev, according to which late Neanderthals continued to exist in small numbers from one end of Asia to the other. Furthermore, as evidenced in historical documents, it showed that the species *Homo neanderthalensis*, which is in any case very polymorphic, had survived in Europe much longer than had been believed, and where, in historic times, it could have given rise here and there to countless stories of wild hairy men, including those of the Scandinavian trolls.

The slow progress forward of this exemplary cryptozoological adventure - from the collection of simple fables widely scattered across all of Eurasia to the accumulation of contemporary eyewitness observations, and crowned by a study of a type specimen - has been recounted in detail in the work which Porshnev and I published in collaboration in 1974, *L'Homme de Neanderthal est Toujours Vivant* (Neanderthal Man is Still Living). Once again, the new discipline had proven that cryptozoology can hasten the discovery of unknown animals.

CRYPTOZOOLOGY HAS ALSO HAD ITS FAILURES

Just as President Jefferson was mistaken in attributing the enormous claw of his *Megalonyx*, a large ground sloth, to a gigantic feline of the Indian traditions, so the Italian naturalist Gian Giuseppe Bianconi (1809-1879) was on the wrong track in relating the eggs and skeletal remains of the *Æpyornis* of Madagascar to the fabulous roc-bird of the *Voyages of Sindbad the Sailor*, that is, to an enormous and bloodthirsty bird of prey.

Let us briefly review the facts. Commencing in 1832, there began to be collected in Madagascar numerous fragments of eggshells and even entire eggs of an unprecedented size. The largest ones measured 32 to 34 cm in length and 22 to 23 cm in diameter: they were thus more than twice as big around and twice as long as ostrich eggs. It was believed for a time that they could be the eggs of monstrous dinosaurs, but when there were found the first subfossil fragments of the skeleton of the creature which had laid them, the evidence had to be accepted: they came from a gigantic bird. In 1851 Isidore Geoffrey Saint-Hilaire (1805-1861) gave this creature the name of *Æpyornis maximus*, which means 'the greatest of the tall birds'.

Here, of course, is something which recalled to a few naturalists certain Arabian and Persian tales which they had read in their youth, such as the stories of *A Thousand and One Nights* and, more especially, those in *The Voyages of Sindbad*. In the second of these the fabulous sailor, having through an oversight been left behind on an island covered with luxurious vegetation, soon discovers there an egg of prodigious size, which 'could have been fifty paces around'. He finally sees the fowl itself, of astounding size, which had just been sitting on it. His companions in the crew had often spoken to him of this fearful bird, which they called a roc. It was capable of the most astonishing exploits. For example, when a rhinoceros disemboweled an elephant, it 'would pick both of them up in its claws and carry them off to feed its little ones'.

More erudite naturalists then recalled that, in the accounts of his travels in the thirteenth century, the Venetian navigator Marco Polo had reported that on certain islands '*oultre Meidagascar sur la côte du Midy*' ('beyond Madagascar to the south') there lived *grifs*, that is, griffins, similar to colossal eagles, thirty paces in size, which were strong enough to pick up an elephant from the ground. On Madagascar itself, these griffins were given the name of roc.

127

Nothing more was needed to suggest to a wonderstruck world that the *Æpyornis*, the existence of which had just been proven, was nothing other than the roc-bird of the Near-Eastern legend. However, Madagascar had never held either elephants or rhinoceroses, and Marco Polo moreover had situated his griffins in the islands located beyond 'Meidagascar'. Finally, it was not even certain that all which he called by this last name was our present Madagascar: some, in fact, have thought instead of Mogadiscio, in Somalia. There was no help for it. The equivalence of roc and *Æpyornis* entered forever into the domain of received ideas, and it is well known that it is practically impossible to wipe out one of these.

All of that would not have been so serious if this received idea had not become a fixation in the mind of Professor Bianconi, of the Royal Academy of Bologna. Between 1862 and 1878, he published no less than eighteen memoirs to show that the bones which had been discovered of the feet of the *Æpyornis* belonged, not to a ratite (that is, a running bird, like the ostrich, the cassowary or the celebrated moas), but to a bird of prey similar to the condors. In 1913, the French paleontologist Louis Monnier summed up quite well the truly heroic undertaking of his Italian colleague: 'Although starting from a preconceived idea and going down the wrong road, he nonetheless very carefully pointed out several interesting anatomical particularities: thickness and shortness of the femur, very unequal condyles, powerful muscular insertions, cervical vertebrae broader than long, characteristics which do not exist in other running birds and which appear at first glance to confirm the hypothesis that it could be a vulturid. When the tibia was discovered, Bianconi, a little taken aback, nevertheless remarked that it indicated an excessive development of the extensor muscles as opposed to atrophy of the flexors, these two systems being equal in true running birds, in order to facilitate walking. This predominance of the extensors in the *Æpyornis* was intended to facilitate the leap prior to taking flight.'

If today it plainly appears absurd that one could even have suspected that the largest of the known birds (it has been estimated that its weight could reach 440 kilos) was capable of flight, it must not be forgotten that, in the last century, the only pieces of evidence available were skimpy fragments of its feet and pelvis. In 1903, Guillaume Grandidier (born in 1873), the son of the great explorer of Madagascar, Alfred Grandidier (1836-1921), was able to write: 'we have today an exact idea of what the *Æpyornis* was, although it has not yet been possible to reconstruct an absolutely complete skeleton of one and the same animal.'

128

The commentary of Dr Monnier showed quite clearly that the undertaking of Bianconi - of cryptozoological inspiration - had really nothing of the ridiculous about it but, on the contrary, exhibited exceptional anatomical knowledge. The great Italian naturalist died in 1879 before having been able to convince himself of his failure. But there are failures which do honour.

THE 'NEW WAVE' OF CRYPTOZOOLOGY

Since the dawn of the twentieth century, and above all because of the sensation created by the discovery of the okapi, the search for which was based on a simple rumour but was carried forward intensively, with diligence and tenacity, a large number of commentators on natural history and even some professional zoologists have begun to devote articles or portions of books to mysterious animals which appeared to be waiting to be discovered.

There should be mentioned here some French popular science writers, like Henri Eugène Victor Coupin (born in 1868), for his book *Les Animaux Excentriques* (Eccentric Animals) (1904); the journalist-reporter Victor Forbin (born in 1864), for an article in *Sciences et Voyages* (Science and Travel) on the possible survival of certain fossils (1920), condensed in 1922 for *Je sais tout* (I know everything); the founder of French speleology Norbert Casteret (1897), and his wife Elisabeth, for their series of articles on 'immemorial' animals, likewise published in *Sciences et Voyages* in 1926; and, finally, Louis Marcellin, one of the erudite pillars of the *Chasseur Français* (French Hunter) of Saint-Etienne, for an article on still unknown animals, which appeared in *Sciences et Voyages* in 1949 and which followed very much the entire affair with great thoroughness. In Great Britain the subject was 'covered' by a whole series of gentlemen from the most varied spheres of activity: the big-game hunter and firearms specialist Walter Winans (1852-1920); an enthusiastic promotor of athleticism, Captain Frederick Annesley Michael Webster (born in 1886); Captain William Lionel Hichens (1874-1940), who had belonged to the Information and Administration Services of East Africa and who, in consequence, had at first hidden behind the pseudonym of 'Fulahn' for his writings on mysterious animals, in 1927 and 1937; and, finally, that trio of prolific authors - all faithful disciples of the great American anomalist Charles Hoy Fort (1874-1932) - Harold Tom Wilkins (1891-1960), Alfred Gordon Bennet (1901-1962) and Eric Frank Russell (1905-1978).

In this lot there must also not be forgotten a few true naturalists who were interested in the matter: the French-Canadian Henry Tilmans, who was of Belgian origin and also went by the name of Henry Tielemans, for his articles on *Raretés Zoologiques* (Zoological Rarities), published from 1905 on in the *Naturaliste Canadien* (Canadian Naturalist); the Catalonian Rossend Serra i Pagès (born in 1863) for his 1923 study entitled *Zoologia Fantastica* (Fantastic Zoology); the French ornithologist Louis Lavauden for his research, published in 1931, on the identity of the legendary animals of Madagascar; the best popular writer in the field of natural science in France, René Thévenin (1877-1967), for all of his writings published widely in the 1930s on still unknown animals; and, finally, the Australian mineralogist Charles Anderson (1876-1944), director of the Australian Museum, for his 1934 article devoted to what he called *The Sea Serpent and its Kind*.

It must also be noted that, in 1917, Laurence Marcellus Larson, the careful and discerning editor of the English translation of the *Konnungs-Skuggsa* (Royal Mirror), of the medieval literature of Norway, took pains to emphasise that exotic marvels are not necessarily false or impossible. And it was in 1931 that the famous German wild-animal collector Joseph Delmont (1875-1935), after having travelled the entire world for some twenty years, wrote the following memorable phrase, which is given considerable weight by his personal experience: 'Without doubt there exist in the regions of the world which are still inaccessible, as well as in the great depths of the sea, creatures which will long remain unknown to science'.

Among the wave of authors of the first half of the twentieth century who were inspired by the problem of hidden animals, it is appropriate, to be sure, to set aside a privileged place for those scientific writers who, in certain chapters of their books, endeavoured to produce a synthesis of the whole of the question, or at least of one of its major aspects.

In Great Britain, these were Frank Walter Lane (born in 1902), for the numerous articles published here and there, which he finally condensed into a chapter which has become a classic in cryptozoology of his book *Nature Parade*, often revised and expanded from 1939 to 1946; the greatest English popular writer in zoology, Dr Maurice Burton (1898-1992), for numerous articles in the *Illustrated London News* and his capital works on animal legends and living fossils, published in 1954, 1955 and 1959, while he was an attaché of the Department of Natural Sciences of the British

Museum; and, finally, the writer Richard Carrington (born in 1921), for his superbly well-documented book *Mermaids and Mastodons* (1957). A book for young people by Richard Ogle, and very well illustrated by its author, *Animals Strange and Rare* (1951) also merits mention, even though the stories told are secondhand.

In Germany, there must first of all be mentioned Dr Richard Hennig (1875-1934), expert geographer and specialist in the history of exploration, for the breadth of view and openness of spirit of his well-reputed work *Wo lag das Paradies?* (Where did Paradise lie?) (1950) and, in particular, the treatment which he gave sea and lake 'monsters'. But again, beyond the Rhine, it was above all the personality of Dr Ingo Krumbiegel who, from 1943 onward, was going to dominate the cryptozoological scene with his various writings. We shall have occasion to come back to him later on, especially because of the primordial importance of his role.

In the United States, the 'avant-garde' of cryptozoology was no less prestigious during the first half of the century. It was made up of three exceptional personalities: the tireless naturalist-traveller Alpheus Hyatt Verrill (1871-1954) (the very son of Professor Addison E. Verrill, to whom we owe the essentials of our knowledge of the super-giant squid), for the openness of mind which characterises all of his popular works in zoology and ethnography, and for the content itself of *Strange Prehistoric Animals* (1948); the Scottish naturalist and writer Ivan Terence Sanderson (1911-1973), based in America since the Second World War, an author full of zest and humour, sometimes inspired with flashes of genius, but – alas! - also a braggart and congenital liar, for a plethora of brilliant articles published between 1947 and 1959; and, above all others, the popular-science writer of German origin, Willy Ley (1906-1969), known especially to the public for his books on astronautics, but who had studied paleontology and the history of zoology at the Universities of Berlin and Koenigsberg, which fact explains the essential character of the works in which he dealt largely with cryptozoology: *The Lungfish, the Dodo and the Unicorn* (1948), *Dragons in Amber* (1951) and *Salamanders and Other Wonders* (1955), as well as *Exotic Zoology* (1959), which is an anthology of the preceding works.

Let us emphasise in passing that, apart from a sort of touristic guide devoted to the eastern part of British Columbia (Canada), *Milestones on the Mighty Fraser* (1950), by Chester Peter Lyons (born in 1915), it was

131

Commander Rupert T. Gould (1890-1948), photo B.B.C.

the Canadian polygraph Richard Stanton Lambert (1894-1981) who was the first to mention the hairy giant of North America, today called Bigfoot, in a book in which he gave it the vernacular Indian name of Sasquatch; this was the book *Exploring the Supernatural: the Weird in Canadian Folklore* (1954). Prior to this, this gigantic primate had been mentioned only in the popular press or, in more or less evasive fashion, in accounts of travel or in ethnographic treatises dealing with native myths.

Certain works by brilliant laymen, reporting on research work in the field, deserve equally to be mentioned for the same period as being first-hand contributions to cryptozoology: the two books, of a rigour beyond reproach, by one of the most sparkling minds of England, Commander Rupert Thomas Gould (1890-1948), on the sea-serpent, *The Case for the Sea-Serpent* (1930) and on lake monsters, *The Loch Ness Monster and Others* (1934); the recollections of Kenneth Cecil Gandar Dower (1908-1944) on his search for the spotted lion of Kenya, *The Spotted Lion* (1937); the report by the hunter and animal painter Moritz Pathe on his pursuit in Liberia of a still unknown spotted antelope, *Die Suche nach dem Fabeltier* (The Search for the Fabulous Animal) (1940); the extensive reports prepared by Ralph Izzard on his hunt for the buru in Assam, *The Hunt for the Buru* (1951), and that for the yeti in Nepal, *The Abominable Snowman Adventure* (1955); then finally, the brief review of aquatic monsters by the Danish museum curator C. M. Poulsen, *Sølangens Gåde* (The Enigma of the Sea Serpent) (1959), supervised and introduced by a first-rate oceanographer, Dr Anton Frederick Bruun (1901-1961).

FROM SOME PINPOINT ATTEMPTS TO A FIRST SUMMING-UP

Oddly enough, none of all of these authors, even the best versed in natural history, ever thought of classifying zoologically and with some degree of precision the mysterious animals about which they had written. Once again, we are dealing here with cryptozoography rather than with cryptozoology. So, especial mention should be made of certain efforts at identification which had been made in the course of the first half of the century by some particularly bold naturalists and zoologists.

Carl Hagenbeck (1844-1913) was not exactly a man of science, but as the 'king of the zoos', the great German animal dealer was exceptionally well-acquainted with the megafauna of the world. A vast network of explorers, collectors of animals and field naturalists kept him informed about even the slightest rumours which circulated here and there concerning apparently new animal forms. He was a merchant, not a dreamer. Thus, he had to be considered as a perfectly reliable authority. That is why he impressed all the men of science as well as the public at large when he disclosed in his memoirs, *Von Tieren und Menschen* (On Animals and Men) (1909), that he had received reports from several quite independent sources concerning the existence, in the immense swamps of tropical Africa, of an unknown animal, 'half-elephant and half-dragon', and especially so when he added: 'According to what I have been able to learn about it, it can only be some species of brontosaur'.

In 1913, the British ornithologist Frank Finn (1868-1938) had been so struck by the recent discovery of the marsupial mole *Notoryctes*, the okapi and the giant forest hog *Hylochoerus* that he wondered if a pygmy water elephant, the existence of which had just been reported in Central Africa, might not instead be a local tapir.

In 1926, the two most eminent mammalogists of Australia, Albert Sherbourne Le Souef (1877-1951) and Harry James Burrell (1873-1945), included in their manual *The Wild Animals of Australasia* a rather large marsupial feline from the north of Queensland, solely on the strength of various reports of encounters. This initiative was further confirmed by Ellis Le Geyt Troughton (1893-1974), curator of mammals of the Australian Museum, in the various successive editions of his classic work *Furred Animals of Australia* (1941, 1954, etc.).

In 1929, the prestigious German naturalist and writer Wilhelm Bölsche (1861-1939) published a small booklet entitled *Drachen, Sage und Naturwissenschaft* (Dragons: Legend and Natural Science), in which he endeavoured to show that dragons had been inspired by the survival into our times of great reptiles from the Mesozoic, such as the dinosaurs. The strongest point in his favour was the scientific report of the Likouala-Congo expedition of 1913-14, led by Captain Freiherr von Stein zu Lausnitz. In this document, the German officer had spoken with a wealth of detail about the mokélé-mbêmbé, an amphibious animal as big as an elephant but having a long neck and a long tail, which had been described by the villagers living along the lower Ubangi and the upper Sanga rivers.

From a snakelike form with two wings, which are doubtless flippers, popular imagination in the Middle Ages created dragons with two wings, with two paws, with two wings and two paws, and finally with two wings and four paws.

In 1933, a school principal of Bozen (today Bolzano) in northern Italy, Dr Jakob Nicolussi, suggested in *Der Schlern*, a local magazine published in German, that the fabulous and fearful tatzelwurm - a stumpy reptile or amphibian, about one meter in length, often observed in the Swiss, Bavarian and Austrian Alps - could be a European species of the only genus of venomous lizard known, namely, *Heloderma*, which is known to exist only in America and is represented by the beaded lizard of Texas, Arizona and the Mexican state of Sonora, otherwise known as the gila monster (*H. suspectum*), and by the beaded lizard of Mexico, otherwise known as the scorpion (*H. horridum*).

In 1934, the British entomologist Malcolm Burr (1878-1954) concluded from an extensive analysis of the literature devoted to the sea-serpent that this creature should be classified among the amphibians, along with the newts and the salamanders.

In 1935, Louis Seymour Bazett Leakey (1903-1972) - who was to become famous one day for the discovery, in Kenya, of the zinjanthropus fossil, nicknamed 'the nutcracker man' - tried to explain the stories circulating in East Africa about a bear of unparalleled ferocity, known to the British colonists under the name of Nandi Bear, by the survival into our time of the *Chalicotherium*. It so happened, in fact, that this strange ungulate with powerful claws, and as big as a horse, had been contemporaneous with the okapi during all of the Miocene and the Pliocene, and thus, like it, could very well have survived in the same forest habitat.

In 1944, Mervyn David Waldegrave Jeffreys (1890-1957), professor at the University of the Witwatersrand in Johannesburg, reviewed critically all of the tales and traditions concerning mysterious 'flying dragons', and summed up his opinion by saying that 'the suspicion lingers that perhaps in some hidden corner of Africa a few shy pterodactyls still lurk'.

In 1945, Dr. William Charles Osman Hill (1901-1978), who was going to become one of the leaders of contemporary primatology, wrote an admirably well-documented study on the long-haired pygmies of Ceylon (today Sri Lanka) the *nittaewo*, exterminated at the end of the eighteenth century, but perhaps represented living today in Sumatra, in the person of the *orang pendek* (little man) or *sedapa*. According to Osman Hill, all of these long-haired pygmies could be late relatives of the fossil ape-man of

Java, at first called *Pithecanthropus* and classified today under the name of *Homo erectus*.

In 1947, Dr Ingo Krumbiegel (1903-1990), the German physician, zoologist and writer, devoted a penetrating study to an amphibious monster of Angola, called locally *coje ya menia* (water lion). From the footprints which it left and the terrible wounds it inflicted on hippopotamuses, Krumbiegel concluded that the animal had to be either a large sabre-toothed feline or a gigantic monitor lizard, like the Komodo dragon, or even some saurian surviving from the Jurassic.

In 1950 as well, Dr Krumbiegel recalled in another article that the first naturalists to study the fauna of New Zealand had reported the sighting and the discovery of tracks, as well as native traditions relating to an animal resembling an otter and which lived in the lakes and rivers of South Island. The Maoris called it *waitoreke*. The presence of this animal, still undiscovered today, on a distant island having no native mammals except for bats (and dogs and rats introduced by man), had long inclined Wilhelm Bölsche, already mentioned above, to surmise, in his principal work *Entwicklungsgeschichte de Natur* (Story of the Evolution of Nature) (1896), that the waitoreke could be related to the monotremes or to an even older group of proto-mammals.

It was also only in 1950 that someone finally attempted to draw a panoramic overview, even if only a schematic one, of the vast field of still unknown animals. This was none other than Krumbiegel himself, in a small book entitled *Von neuen und unentdeckten Tierarten* (On New and Undiscovered Animal Species). In this work, he first reviewed the major zoological discoveries made in the course of about the past hundred years, and then he enumerated all unidentified animals for which rumours of their existence had come to him. He listed altogether about a dozen: the king cheetah of Southern Rhodesia (today Zimbabwe), the entirely black tapir of Sumatra, the ameranthropoid of Venezuela, the 'water lion' of Angola, the sea-serpent of the Atlantic and the Pacific, the fossil *Chirotherium* of Thuringia, known only from its footprints, the waitoreke of New Zealand, a small spotted bushbuck of West Africa, a new goral, a sort of goat antelope from the north of India, the bipunctate argus, a pheasant known only from a feather, and the marsupial tiger of the north of Queensland. This was doubtless rather little in comparison to the some 150 forms which I was going to enumerate in my *Checklist* of 1986, but it was a beginning, and one rich in promise.

137

Thus, in the middle of the present century there was already extant in the scientific literature not only the exemplary study of Dr. Oudemans on 'the Great Unknown' of the seas but also, among about a dozen more or less well-founded suggestions as to the zoological identity of certain terrestrial animals of this sort, there was an equally exemplary work work by Dr. Osman Hill on one of them, and even a brief overview of some major cryptozoological problems by Dr Krumbiegel.

CRYPTOZOOLOGY BAPTISED AT LAST

It was at this point that I, Dr. Bernard Heuvelmans (born in 1916 in France, of a Belgian father and a Dutch mother), entered the scene as a researcher aiming at exhaustiveness, seeking to achieve a true synthesis of all previous research work, and as 'refiner' of the methodology, until then still only roughed out, of the new zoological discipline.

My cryptozoological spirit had doubtless been inspired during my turbulent childhood by three important novels: *Vingt Mille Lieues sous les Mers* (Twenty Thousand Leagues Under the Sea), by Jules Verne (1828-1905), because of its story of encountering a super-giant squid and various other sea monsters; *The Lost World*, by Arthur Conan Doyle (1859-1938), because it assumed the survival of prehistoric animals on the summit of an isolated plateau in South America; and *Les Dieux Rouges* (The Red Gods), by Jean d'Esme (pseudonym of the Viscount Jean d'Esmenard, born in 1893), because it is based on the existence in Indochina of an unknown tribe of hairy wild men, doubtless ape-men, a story published in 1928 and which anticipated the much better known novel of Vercors (pseudonym of Charles Bruller, 1902-1991), *Les Animaux Denatures* (Unnatural Animals) (1952), but without its philosophical implications. This being said, it was only after having read in the *Saturday Evening Post* of 3 January 1948 a sensational article by Ivan T. Sanderson, *There Could be Dinosaurs*, which dealt with the possible survival of dinosaurs in Africa, that I made the decision to write an exhaustive work covering all cases of this sort. In fact, I had accumulated over the years a substantial quantity of information on this subject, notably while I was studying zoology at the Free University of Brussels, from 1934 to 1939. Clearly, I could not at that time have been aware of the imminent publication of the digest by Krumbiegel, which was going to prove to be most helpful in the course of my work.

Image from Jules Verne's Twenty Thousand Leagues Under the Sea.

I worked steadily for four full years on the execution of my project, which dealt with a variety of cases - several dozen, in fact - in the most prosaic manner and in which, for the first time in a work devoted to such problems, all of my sources of information were cited, in conformity with the rules established for scientific publications. From this labour of love there came first of all a series of preliminary articles, published in 1952 in the popular-science review *Sciences et Avenir* (Science and the Future) and then, finally, a major work in two volumes, limited by design to terrestrial animals only, *Sur la Piste des Bêtes Ignorées* (On the Track of Unknown Animals) (1955). This work was subsequently translated into more than ten foreign

languages, which fact explains the considerable repercussions which it had throughout the world.

However, it was only in working with the problem of unknown animals of the sea that I came to understand that the method of Oudemans was far from being perfect, and that it was hence necessary for it to be seriously expanded and refined. I recognised above all that the one whom I considered with devotion as my master had committed a grave error in considering *a priori* that *all* sea-serpents - in sum, all of the large and elongated marine animals, whether still unknown to science or not - belonged to one and the same species. It soon became apparent to me that some of them could not even belong to the same class of vertebrates. Aside from one large reptile - and certainly not a snake! - and a gigantic turtle, there seemed to be involved not only large pinnipeds, following Oudemans, and archaic cetaceans, following Wood, but a whole host of fish, plainly unrelated to one another. To disentangle successfully this atrociously complex problem, I had finally to resort to the services of the most primitive of computers, the punched-card system.

Over a number of years, I strove to perfect the purely statistical approach of Oudemans, which is effective only for the clarification of a single isolated phenomenon, and this, to be sure, is rarely the case in cryptozoological matters. Finally, I was able to construct a generalised method capable of being applied - without preconceived ideas, without prejudices, without subjective judgements - to all cases of recent information or ancient beliefs concerning apparently unknown animals.

To resolve the problem of the great sea-serpent, I at first studied for three years the virtually parallel question of another sea monster, this one even more fantastic, the kraken, which in any event finally was solved by the discovery of gigantic squids of the genus *Architeuthis*. Further, this analysis had also strengthened my conviction that a unique and reductionist explanation was often not defensible in cryptozoology. Here as well, several different creatures, outsized octopuses, still unknown, as well as enormous squids, henceforth unmasked, had joined together to support the legend of the tentacular island-beast. All of this was the object of the work *Dans le Sillage des Monstres Marins: le Kraken'et le Poulpe Colossal* (In the Wake of Sea Monsters, the Kraken and the Colossal Octopus) (1958). Only after this did I pass seven more years in examining more thoroughly the tangled

case of sea-serpents themselves, all of which was to lead finally to the publication in 1965 of *Le Grand Serpent-de-Mer: le Problème Zoologique et sa Solution* (The Great Sea-Serpent: The Zoological Problem And Its Solution).

It was during these years of intensive work, toward the latter part of the 1950s, that I felt the need to give a name to what his research implied to be a new discipline of science. So, this is how I coined the term 'cryptozoology', meaning 'the science of hidden animals', which I then began to use regularly in my professional correspondence. The word appeared in print for the first time in 1959, when one of my numerous correspondents, Lucien Blancou (1903-1983), Honorary Chief Inspector of Hunting and Protection of Overseas Fauna, dedicated his work *Géographie Cynégétique du Monde* (Cynegetic Geography of the World) '*A Bernard Heuvelmans, maître de la Cryptozoologie*' (To Bernard Heuvelmans, Master of Cryptozoology). This term today is widely used throughout the entire world, and figures as well in numerous dictionaries and encyclopedias. This was the root of the family tree of cryptozoology.

THE EXEMPLARITY OF THE DISCOVERY OF NEPTUNE

It should by now be clear that the essence of cryptozoological research lies in the diligent gathering, the analysis, the sifting through to eliminate mistakes and evident hoaxes, the comparative study and the synthesis of all information available on animals which are still absent from our catalogues, even though we have bundles of information on them. The final result of this long-drawn-out process - a sort of moral as well as physical identikit picture - should in principle enable the creatures in question to be localised with the greatest precision possible and to be recognised unambiguously. Further, it should enable us to know where, when and how to approach them, perhaps to draw them into a trap, and finally to capture them, even if only on film, and, ideally, to establish peaceful relations and even bonds of friendship with them.

To go trekking through the Himalayas in the hope of a chance meeting with a snowman frolicking about, to go diving in Loch Ness to try to harpoon one of the local 'monsters', to track the Bigfoot through the forested western mountains of North America, to go paddling or wading through oppressive tropical swamps in search of dinosaurs which have come down from the

Secondary, doubtless is much more fun, but has nothing in the world to do with cryptozoological research properly speaking; no more than does dredging the bottom of the sea, exploring the dark waters of subterranean caverns, or just hunting butterflies or beetles with a net. Activities of this type are only routine zoological tasks: blind samplings, exploratory tests or processes of control. In fact, true cryptozoological research in the field consists above all of collecting from the local people additional information which is more complete and more recent, and in searching at the same time for possible concrete evidence of the existence of the animals being sought and, at the end of all of this, endeavouring finally to meet up with the creatures under the most favourable conditions.

The ambitious aim of cryptozoology is to succeed in describing an animal scientifically *before* having had to capture it or kill it. This is on the whole a generous aim, which stems from an ethic that is more compatible with the respect due to a nature which today is dangerously threatened.

Such a feat can best be compared to the discovery in 1846 of the planet Neptune by Urbain Le Verrier (1811-1877). It was through the study of a slight perturbation in the elliptic orbit of Uranus that the French astronomer was able to deduce the presence of a hitherto unsuspected planet. Le Verrier informed his German colleague Johann Galle (1812-1910) exactly where to aim his telescope if he wanted to see the unknown planet, and also the luminosity and size which it represented. Two or three days later, Galle discovered Neptune within a degree of the point which Le Verrier had indicated to him.

The celebrated popular-science writer Camille Flammarion (1842-1925) reported that, although Le Verrier had been named director of the Paris Observatory as a result of his brilliant success, he never even took the trouble to glance at the heavenly body which he had discovered. It is doubtful that a cryptozoologist would ever exhibit the same admirable detachment as did this rigorous theoretician, should a specimen of the species described by him happen to be captured or stranded providentially on a beach.

INDEX

The author in front of Gloucester Harbour, Massachusetts, in 1968.